Lecture Notes in Mathematics

Edited by J.-M. Morel, F. Takens and B. Teissier

Editorial Policy
for the publication of monographs

1. Lecture Notes aim to report new developments in all areas of mathematics and their applications – quickly, informally and at a high level. Mathematical texts analysing new developments in modelling and numerical simulation are welcome.

   Monograph manuscripts should be reasonably self-contained and rounded off. Thus they may, and often will, present not only results of the author but also related work by other people. They may be based on specialised lecture courses. Furthermore, the manuscripts should provide sufficient motivation, examples and applications. This clearly distinguishes Lecture Notes from journal articles or technical reports which normally are very concise. Articles intended for a journal but too long to be accepted by most journals, usually do not have this „lecture notes" character. For similar reasons it is unusual for doctoral theses to be accepted for the Lecture Notes series, though habilitation theses may be appropriate.

2. Manuscripts should be submitted (preferably in duplicate) either to Springer's mathematics editorial in Heidelberg, or to one of the series editors (with a copy to Springer). In general, manuscripts will be sent out to 2 external referees for evaluation. If a decision cannot yet be reached on the basis of the first 2 reports, further referees may be contacted: The author will be informed of this. A final decision to publish can be made only on the basis of the complete manuscript, however a refereeing process leading to a preliminary decision can be based on a pre-final or incomplete manuscript. The strict minimum amount of material that will be considered should include a detailed outline describing the planned contents of each chapter, a bibliography and several sample chapters.

   Authors should be aware that incomplete or insufficiently close to final manuscripts almost always result in longer refereeing times and nevertheless unclear referees' recommendations, making further refereeing of a final draft necessary.

   Authors should also be aware that parallel submission of their manuscript to another publisher while under consideration for LNM will in general lead to immediate rejection.

3. Manuscripts should in general be submitted in English. Final manuscripts should contain at least 100 pages of mathematical text and should always include

   – a table of contents;
   – an informative introduction, with adequate motivation and perhaps some historical remarks: it should be accessible to a reader not intimately familiar with the topic treated;
   – a subject index: as a rule this is genuinely helpful for the reader.

   For evaluation purposes, manuscripts may be submitted in print or electronic form (print form is still preferred by most referees), in the latter case preferably as pdf- or zipped ps-files. Lecture Notes volumes are, as a rule, printed digitally from the authors' files. To ensure best results, authors are asked to use the LaTeX2e style files available from Springer's web-server at:

   ftp://ftp.springer.de/pub/tex/latex/mathegl/mono/ (for monographs) and

   ftp://ftp.springer.de/pub/tex/latex/mathegl/mult/ (for summer schools/tutorials).

   Additional technical instructions, if necessary, are available on request from lnm@springer-sbm.com.

Continued on inside back-cover

Lecture Notes in Mathematics 1879

Editors:
J.-M. Morel, Cachan
F. Takens, Groningen
B. Teissier, Paris

Subseries:
Ecole d'Eté de Probabilités de Saint-Flour

G. Slade

The Lace Expansion
and its Applications

Ecole d'Eté de Probabilités
de Saint-Flour XXXIV - 2004

Editor: Jean Picard

 Springer

Author

Gordon Slade
Department of Mathematics
University of British Columbia
121 -1984 Mathematics Road
Vancouver, BC., V6T 1Z2
Canada
e-mail: slade@math.ubc.ca

Editor

Jean Picard
Laboratoire de Mathématiques Appliquées
UMR CNRS 6620
Université Blaise Pascal Clermont-Ferrand
63177 Aubière Cedex
France
e-mail: jean.picard@math.univ-bpclermont.fr

Cover: Blaise Pascal (1623-1662)

Library of Congress Control Number: 2006921789

Mathematics Subject Classification (2000): 60K35, 82B41, 82B43, 60G57, 05A16

ISSN print edition: 0075-8434
ISSN electronic edition: 1617-9692
ISSN Ecole d'Eté de Probabilités de St. Flour, print edition: 0721-5363
ISBN-10 3-540-31189-0 Springer Berlin Heidelberg New York
ISBN-13 978-3-540-31189-8 Springer Berlin Heidelberg New York
DOI 10.1007/978-3-540-35518-2

Springer is a part of Springer Science+Business Media
springer.com
© Springer-Verlag Berlin Heidelberg 2006

Typesetting: by the authors and SPI Publisher Services using a Springer LaTeX package

Cover design: design & production GmbH, Heidelberg

Printed on acid-free paper   SPIN: 11602538   VA41/3100/SPI   5 4 3 2 1 0

Foreword

Three series of lectures were given at the 34th Probability Summer School in Saint-Flour (July 6–24, 2004), by the Professors Cerf, Lyons and Slade. We have decided to publish these courses separately. This volume contains the course of Professor Slade. We cordially thank the author for his performance at the summer school, and for the redaction of these notes.

69 participants have attended this school. 35 of them have given a short lecture. The lists of participants and of short lectures are enclosed at the end of the volume.

The Saint-Flour Probability Summer School was founded in 1971. Here are the references of Springer volumes which have been published prior to this one. All numbers refer to the *Lecture Notes in Mathematics* series, except S-50 which refers to volume 50 of the *Lecture Notes in Statistics* series.

1971: vol 307	1980: vol 929	1990: vol 1527	1998: vol 1738
1973: vol 390	1981: vol 976	1991: vol 1541	1999: vol 1781
1974: vol 480	1982: vol 1097	1992: vol 1581	2000: vol 1816
1975: vol 539	1983: vol 1117	1993: vol 1608	2001: vol 1837 & 1851
1976: vol 598	1984: vol 1180	1994: vol 1648	2002: vol 1840
1977: vol 678	1985/86/87: vol 1362 & S-50	1995: vol 1690	2003: vol 1869
1978: vol 774	1988: vol 1427	1996: vol 1665	2004: vol. 1878 & 1879
1979: vol 876	1989: vol 1464	1997: vol 1717	

Further details can be found on the summer school web site
http://math.univ-bpclermont.fr/stflour/

Jean Picard, Université Blaise Pascal
Chairman of the summer school

Preface

Several superficially simple mathematical models, such as the self-avoiding walk and percolation, are paradigms for the study of critical phenomena in statistical mechanics. Although these models have been studied by mathematicians for about half a century, exciting new developments continue to occur and the subject is flourishing. Much progress has been made, but it remains a major challenge for mathematical physics and probability theory to obtain a complete and mathematically rigorous understanding of the scaling theory of these models at criticality.

These lecture notes concern the lace expansion, which is a powerful tool for the analysis of the critical scaling of several models above their upper critical dimensions, namely:

- the self-avoiding walk on $\mathbb{Z}^d$ for $d > 4$,
- lattice trees and lattice animals on $\mathbb{Z}^d$ for $d > 8$,
- percolation on $\mathbb{Z}^d$ for $d > 6$,
- oriented percolation on $\mathbb{Z}^d \times \mathbb{Z}_+$ and the contact process on $\mathbb{Z}^d$ for $d > 4$.

Results include proofs of existence of critical exponents, with mean-field values, and construction of scaling limits. Often, the scaling limit is described in terms of super-Brownian motion.

There are two distinct goals for these notes. The first goal is to provide a written accompaniment to my lectures at the XXXIV Saint–Flour International Probability School, in July 2004, and at the Pacific Institute for the Mathematical Sciences – University of British Columbia Summer School on Probability, in June 2005. The notes contain an introduction to the lace expansion and several of its applications, with sufficient background and depth to prepare a newcomer to do research using the lace expansion. Basic graduate level probability theory will be used, but no previous knowledge of the lace expansion or super-Brownian motion is assumed. The second goal is to provide a survey of the field, so that an interested reader can follow up by consulting the original literature. In pursuit of the second goal, these notes include more material than can be covered during a summer school course.

Following a brief initial chapter concerning random walk, the notes can be divided into four parts, whose contents are summarized as follows.

Part I, which concerns the self-avoiding walk, consists of Chaps. 2–6. A complete and self-contained proof is given of the convergence of the lace expansion for the nearest-neighbour model in dimensions $d \gg 4$, and for the spread-out model of self-avoiding walks which take steps of length at most L, with $L \gg 1$, in dimensions $d > 4$. The convergence proof presented here seems simpler than all previous lace expansion convergence proofs. As a consequence of convergence, it is shown that the critical exponent γ for the generating function of the number of n-step self-avoiding walks exists and is equal to 1. A survey is then given of the many extensions of this result that have been obtained using the lace expansion.

Part II, which concerns lattice trees and lattice animals, consists of Chaps. 7–8. It is shown how a minor modification of the expansion for the self-avoiding walk can be applied to give expansions for lattice trees and lattice animals, and an indication is given of the diagrammatic estimates that are necessary for proving convergence of the expansion. The relevance of the square condition is indicated, and results concerning existence of critical exponents in dimensions $d > 8$ are surveyed.

Part III, which concerns percolation, oriented percolation, and the contact process, consists of Chaps. 9–14. Detailed discussions are given of expansions for each of these models. Differential inequalities involving the triangle condition are stated (and usually proved) and are shown to imply mean-field behaviour of various critical exponents. Results concerning existence of critical exponents in dimensions $d > 6$ (for percolation) and $d > 4$ (for oriented percolation and the contact process) are surveyed.

Part IV, which concerns super-Brownian scaling limits, consists of Chaps. 15–17. Critical branching random walk with Poisson offspring distribution is analyzed in detail and used to give a self-contained construction of integrated super-Brownian excursion (ISE). The role of ISE as the scaling limit of lattice trees and of critical percolation clusters, above the upper critical dimensions, is discussed. The canonical measure of super-Brownian motion is also described, as is its role as scaling limit of critical oriented percolation clusters and the critical contact process in dimensions $d > 4$, and of lattice trees in dimensions $d > 8$.

Mathematics is not a spectator sport, and true understanding requires active participation in working out the ideas. To help facilitate this, a number of exercises for the reader appear throughout the notes. Some can be solved in a few lines, and others require more effort. I am grateful to Jeremy Flowers, Jesse Goodman, Jeffrey Hood, Sandra Kliem, Richard Liang, and Terry Soo, who collectively wrote solutions to all the exercises during the PIMS–UBC summer school.

It would not be possible to include detailed proofs of all the results discussed in these lecture notes without substantially increasing their length, and a number of important topics are only alluded to. These include: the inductive approach to the lace expansion, which is in many respects the most powerful method to prove convergence of the expansion; the "double" expansions that have been used to analyze r-point functions for $r \geq 3$; and the lace expansion on a tree, which is a method that can sometimes be used to replace a double expansion. (Two of these topics—the inductive method and double expansions—are discussed in recent lecture notes by Remco van der Hofstad [110].) Also, a complete proof of the convergence of the expansion is given only for the self-avoiding walk. This is the simplest setting for proving convergence, and convergence for the other models can be based on the ideas used in this setting. Finally, in an important new development about which it is too early to provide details, Sakai [181] has shown how to apply the lace expansion to analyze the Ising model in dimensions $d > 4$.

This work was supported in part by NSERC of Canada. Versions of the lectures were given at the University of British Columbia in Spring 2003, at EURANDOM in Fall 2003, at Saint-Flour in Summer 2004, and at PIMS/UBC in Summer 2005. The lecture notes were written primarily while I was travelling during 2003-04. I thank EURANDOM and the Thomas Stieltjes Institute, the University of Melbourne, Microsoft Research, and my hosts at these institutions, for their hospitality during visits to Eindhoven, Melbourne and Redmond.

I am grateful to the friends and colleagues with whom I have had the good fortune to work on topics related to these lecture notes. I thank Markus Heydenreich, Remco van der Hofstad, Mark Holmes, Sandra Kliem, Ed Perkins and Akira Sakai for suggesting improvements and for comments on earlier drafts of these notes. Many others have also made helpful comments of one form or another. Most of the illustrations (and all of the best ones) were produced by Bill Casselman, my colleague at the University of British Columbia and Graphics Editor of *Notices of the American Mathematical Society*.

I extend special thanks to David Brydges, whose patient teaching brought me into the subject, and to Takashi Hara and Remco van der Hofstad, who have played profound roles in the development of the ideas presented in these notes.

Vancouver,   Gordon Slade
August 9, 2005

Contents

1

Simple Random Walk

The point of departure for the lace expansion is simple (ordinary) random walk, and it is helpful first to recall some elementary facts about random walk on $\mathbb{Z}^d$. This will also set some notation for later use.

1.1 Asymptotic Behaviour

Fix a finite set $\Omega \subset \mathbb{Z}^d$ that is invariant under the symmetry group of $\mathbb{Z}^d$, i.e., under permutation of coordinates or replacement of any coordinate x_i by $-x_i$. Our two basic examples are the *nearest-neighbour model*

$$\Omega = \{x \in \mathbb{Z}^d : \|x\|_1 = 1\} \tag{1.1}$$

and the *spread-out model*

$$\Omega = \{x \in \mathbb{Z}^d : 0 < \|x\|_\infty \leq L\}, \tag{1.2}$$

where L is a fixed (usually large) constant. The norms are defined, for $x = (x_1, \ldots, x_d)$, by $\|x\|_1 = \sum_{j=1}^d |x_j|$ and $\|x\|_\infty = \max_{1 \leq j \leq d} |x_j|$.

For $n \geq 1$, an n-step walk taking steps in Ω is defined to be a sequence $(\omega(0), \omega(1), \ldots, \omega(n))$ of vertices in $\mathbb{Z}^d$ such that $\omega(i) - \omega(i-1) \in \Omega$ for $i = 1, \ldots, n$. Let $\mathcal{W}_n(x, y)$ be the set of n-step walks with $\omega(0) = x$ and $\omega(n) = y$, and let $\mathcal{W}_n = \cup_{x \in \mathbb{Z}^d} \mathcal{W}_n(0, x)$ denote the set of all n-step walks starting from the origin. Let $c_n^{(0)}(x)$ denote the cardinality of $\mathcal{W}_n(0, x)$. The superscript (0) is there to indicate that we are working with the random walk with no interaction. We allow for the degenerate case $n = 0$ by defining $\mathcal{W}_0(x, y)$ to consist of the zero-step walk (x) if $x = y$, and to be empty otherwise. Then $c_0^{(0)}(x, y) = \delta_{x,y}$. Taking into account the translation invariance, we will use the abbreviations $\mathcal{W}_n(y - x) = \mathcal{W}_n(x, y)$ and $c_n^{(0)}(y - x) = c_n^{(0)}(x, y)$.

For $n \geq 1$, by considering the possible values $y \in \Omega$ of the walk's first step, we have

$$c_n^{(0)}(x) = \sum_{y \in \Omega} c_{n-1}^{(0)}(x-y) = \sum_{y \in \mathbb{Z}^d} c_1^{(0)}(y)c_{n-1}^{(0)}(x-y). \qquad (1.3)$$

Denoting the convolution of functions f and g by

$$(f * g)(x) = \sum_{y \in \mathbb{Z}^d} f(y)g(x-y), \qquad (1.4)$$

(1.3) can be written as

$$c_n^{(0)}(x) = \left(c_1^{(0)} * c_{n-1}^{(0)}\right)(x). \qquad (1.5)$$

The Fourier transform of an absolutely summable function $f : \mathbb{Z}^d \to \mathbb{C}$ is defined by

$$\hat{f}(k) = \sum_{x \in \mathbb{Z}^d} f(x)e^{ik \cdot x} \qquad (k \in [-\pi, \pi]^d), \qquad (1.6)$$

where $k \cdot x = \sum_{j=1}^d k_j x_j$, with inverse

$$f(x) = \int_{[-\pi,\pi]^d} \frac{d^d k}{(2\pi)^d} \hat{f}(k)e^{-ik \cdot x}. \qquad (1.7)$$

The fact stated in part (a) of the following exercise makes the use of Fourier transforms very convenient.

Exercise 1.1. (a) Show that the Fourier transform of $f * g$ is $\hat{f}\hat{g}$.
(b) A closely related statement is the following. Denote the generating functions of the sequences f_n and g_n by $F(z) = \sum_{n=0}^{\infty} f_n z^n$ and $G(z) = \sum_{n=0}^{\infty} g_n z^n$, and assume these series both have positive radius of convergence. Show that the generating function $H(z)$ of the sequence $h_n = \sum_{m=0}^{n} f_m g_{n-m}$ is $H(z) = F(z)G(z)$.

By Exercise 1.1(a), (1.5) implies that

$$\hat{c}_n^{(0)}(k) = \hat{c}_1^{(0)}(k)\hat{c}_{n-1}^{(0)}(k). \qquad (1.8)$$

Since $\hat{c}_0^{(0)}(k) = 1$, solving (1.8) by iteration gives

$$\hat{c}_n^{(0)}(k) = \hat{c}_1^{(0)}(k)^n \qquad (n \geq 0). \qquad (1.9)$$

If we define the transition probability

$$D(x) = \frac{1}{|\Omega|}I[x \in \Omega] = \frac{1}{|\Omega|}c_1^{(0)}(x), \qquad (1.10)$$

where $|\Omega|$ denotes the cardinality of the set Ω and I denotes the indicator function, then (1.9) can be rewritten as

$$\hat{c}_n^{(0)}(k) = |\Omega|^n \hat{D}(k)^n \qquad (n \geq 0). \qquad (1.11)$$

Exercise 1.2. (a) Show that for the nearest-neighbour model,

$$\hat{D}(k) = \frac{1}{d} \sum_{j=1}^{d} \cos k_j, \tag{1.12}$$

and for the spread-out model

$$\hat{D}(k) = \frac{1}{|\Omega|} \left[\prod_{j=1}^{d} \hat{M}(k_j) - 1 \right], \tag{1.13}$$

where

$$M(t) = \frac{\sin[(2L+1)t/2]}{\sin(t/2)} \tag{1.14}$$

is the Dirichlet kernel.

(b) Denote the variance of D by $\sigma^2 = \sum_{x \in \mathbb{Z}^d} |x|^2 D(x)$. Show that $\sigma = 1$ for the nearest-neighbour model and that σ is asymptotic to a multiple of L as $L \to \infty$ for the spread-out model.

The number of n-step walks starting from a given vertex is of course $|\Omega|^n$, because each step can be chosen in $|\Omega|$ different ways. This fact is contained in (1.11), since the number of n-step walks starting from the origin is $\sum_{x \in \mathbb{Z}^d} c_n^{(0)}(x) = \hat{c}_n^{(0)}(0) = |\Omega|^n$, using $\hat{D}(0) = 1$.

By symmetry, $\sigma^2 = -\nabla^2 \hat{D}|_{k=0}$, where $\nabla^2 = \sum_{j=1}^{d} \nabla_j^2$ is the Laplacian, with ∇_j denoting partial differentiation with respect to the component k_j of k. Then, by (1.11) and by the symmetry of Ω, the central limit theorem

$$\lim_{n \to \infty} \frac{\hat{c}_n^{(0)}(k/\sigma\sqrt{n})}{\hat{c}_n^{(0)}(0)} = e^{-|k|^2/2d} \tag{1.15}$$

follows, as does the fact that the mean-square displacement is given by

$$\frac{\sum_{x \in \mathbb{Z}^d} |x|^2 c_n^{(0)}(x)}{\sum_{x \in \mathbb{Z}^d} c_n^{(0)}(x)} = -\nabla^2 \hat{D}^n \bigg|_{k=0} = n\sigma^2. \tag{1.16}$$

Exercise 1.3. Prove (1.15) and (1.16).

The *two-point function* is defined by

$$C_z(x,y) = \sum_{n=0}^{\infty} \sum_{\omega \in \mathcal{W}_n(x,y)} z^n = \sum_{n=0}^{\infty} c_n^{(0)}(x,y) z^n. \tag{1.17}$$

The two-point function is finite for $z \in [0, 1/|\Omega|)$. For $d > 2$, it is also known to be finite for $z = 1/|\Omega|$, and for this value of z it is called the Green function. By translation invariance, we may regard the two-point function as a function of a single variable, writing $C_z(x,y) = C_z(y-x)$. By (1.11) and (1.17), its Fourier transform is

$$\hat{C}_z(k) = \sum_{n=0}^{\infty} \hat{c}_n^{(0)}(k)z^n = \frac{1}{1 - z|\Omega|\hat{D}(k)}. \tag{1.18}$$

The *susceptibility* is defined by

$$\chi(z) = \sum_{x\in\mathbb{Z}^d} C_z(0,x) = \hat{C}_z(0) = \frac{1}{1 - z|\Omega|}. \tag{1.19}$$

The *critical point* is the singularity $z_c = 1/|\Omega|$ of the susceptibility.

The inverse Fourier transform of (1.18) is

$$C_z(x) = \int_{[-\pi,\pi]^d} \frac{d^d k}{(2\pi)^d} \frac{e^{-ik\cdot x}}{1 - z|\Omega|\hat{D}(k)}. \tag{1.20}$$

For $d > 2$,

$$C_{z_c}(x) \sim \text{const} \frac{1}{|x|^{d-2}} \tag{1.21}$$

as $|x| \to \infty$, where the constant depends on d and on Ω (see [149,195], or [203] for a more general statement of this fact). The notation

$$f(x) \sim g(x) \qquad \text{denotes} \qquad \lim_{x\to\infty} f(x)/g(x) = 1, \tag{1.22}$$

and this notation will used in general for asymptotic formulas.

Exercise 1.4. Some care is needed with (1.20) when $z = z_c$, since $C_{z_c}(x)$ is not summable by (1.21) and thus its Fourier transform is problematic. Using the symmetry of Ω, prove that (1.20) does hold when $z = z_c$ for $d > 2$, and that the integral is infinite when $z = z_c$ for $d \le 2$.

Exercise 1.5. Let $f : \mathbb{Z}^d \to \mathbb{C}$. For $y \in \Omega$, define forward and backward discrete partial derivatives by $\partial_y^+ f(x) = f(x+y) - f(x)$ and $\partial_y^- f(x) = f(x) - f(x-y)$. Define the discrete Laplacian by

$$\Delta f(x) = \frac{1}{2}\frac{1}{|\Omega|}\sum_{y\in\Omega}\partial_y^-\partial_y^+ f(x) = \frac{1}{|\Omega|}\sum_{y\in\Omega} f(x+y) - f(x), \tag{1.23}$$

and let $\delta_{x,y}$ denote the Kronecker delta which takes the value 1 if $x = y$ and 0 if $x \ne y$. Show that $-\Delta C_{1/|\Omega|}(x) = \delta_{0,x}$. Thus $C_{1/|\Omega|}(x)$ is the Green function for $-\Delta$.

Exercise 1.6. Consider a simple random walk started at the origin.
(a) Let u denote the probability that the walk ever returns to the origin. The walk is *recurrent* if $u = 1$ and *transient* if $u < 1$. Let N denote the (random) number of visits to the origin, including the initial visit at time 0, and let

$m = \mathbb{E}N$. Show that $m = \frac{1}{1-u}$, so the walk is recurrent if and only if $m = \infty$.
(b) Show that

$$m = \sum_{n=0}^{\infty} \mathbb{P}(\omega(n) = 0) = \int_{[-\pi,\pi]^d} \frac{1}{1 - \hat{D}(k)} \frac{\mathrm{d}^d k}{(2\pi)^d}. \tag{1.24}$$

Thus transience is characterized by the integrability of $\hat{C}_{1/|\Omega|}(k)$.
(c) For simplicity, consider the nearest-neighbour model, with Ω given by
(1.1). Show that the walk is recurrent in dimensions $d \leq 2$ and transient in
dimensions $d > 2$.

Exercise 1.7. Let $\omega^{(1)}$ and $\omega^{(2)}$ denote two independent simple random walks
started at the origin, and let

$$X = \sum_{i=0}^{\infty} \sum_{j=0}^{\infty} I[\omega^{(1)}(i) = \omega^{(2)}(j)] \tag{1.25}$$

denote the number of intersections of the two walks. Here I denotes an indicator function. Show that

$$\mathbb{E}X = \int_{[-\pi,\pi]} \frac{1}{[1 - \hat{D}(k)]^2} \frac{\mathrm{d}^d k}{(2\pi)^d}. \tag{1.26}$$

Thus $\mathbb{E}X$ is finite if and only if $\hat{C}_{1/|\Omega|}(k)$ is square integrable. Conclude,
for simplicity for the nearest-neighbour model, that the expected number of
intersections is finite if $d > 4$ and infinite if $d \leq 4$.

The integral $(2\pi)^{-d} \int_{[-\pi,\pi]^d} \hat{C}_{z_c}(k)^2 \mathrm{d}^d k$ of (1.26) is equal, by the Parseval
relation, to $\sum_{x \in \mathbb{Z}^d} C_{z_c}(x)^2$. The relevance of the condition $d > 4$ for the
latter is evident from the asymptotic behaviour (1.21). However, the k-space
analysis is more elementary, as it relies on the easy formulas given in (1.12)
and (1.18) rather than the deeper statement (1.21). It is often much easier to
use estimates in k-space than to work directly in x-space.

It is a consequence of Donsker's Theorem [24] that the scaling limit of
simple random walk is Brownian motion, in all dimensions. This means that
if we define a random continuous function X_n from the interval $[0,1]$ into $\mathbb{R}^d$
by setting $X_n(j/n) = \sigma^{-1} n^{-1/2} \omega(j)$ for integers $j \in [0,n]$, and interpolating
linearly between consecutive vertices, then the distribution of X_n converges
weakly to the Wiener measure. See Fig. 1.1.

1.2 Universality and Spread-Out Models

In these notes, we study several models that live on the integer lattice,
and each has a nearest-neighbour and a spread-out version. In the nearest-
neighbour model, specified by (1.1), bonds (also called edges) join pairs of

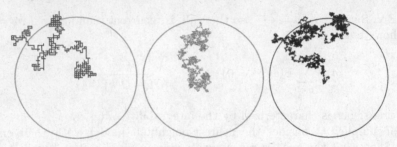

Fig. 1.1. Nearest-neighbour random walks on $\mathbb{Z}^2$ taking $n = 1,000$, $10,000$ and $100,000$ steps. The circles have radius $\sqrt{n}$, in units of the step size of the random walk.

vertices separated by unit Euclidean distance. In the spread-out model, specified by (1.2), bonds join pairs of vertices separated by distance between 1 and L, where L is a parameter usually taken to be large. According to the deep hypothesis of *universality*, the critical scaling of the models to be studied should be the same for the nearest-neighbour and spread-out models.

We use the spread-out model because proofs of convergence of the lace expansion require *large degree*. The degree is the cardinality of Ω. For the nearest-neighbour model the degree is $2d$, and can be taken large by increasing the dimension. The degree of the spread-out model is of order L^d for large L, and this allows for convergence proofs for the lace expansion without taking the dimension d to be large in an uncontrolled way. In the applications to be discussed, results will typically be obtained: (i) for the nearest-neighbour model for $d \geq d_0$ for some d_0 having no physical meaning, and (ii) for the spread-out model with L larger than some L_0 and d strictly greater than the upper critical dimension (4 for the self-avoiding walk, oriented percolation and the contact process; 6 for percolation; 8 for lattice trees and lattice animals). While it is of interest to prove results of type (i) with d_0 equal to the upper critical dimension plus one, failing this, results of type (ii) seem more important, as they indicate clearly the role of the upper critical dimension. Also, the fact that all large L give rise to the same scaling behaviour provides a partial proof of universality in this context. In fact, much more general spread-out models than (1.2) can be handled using the lace expansion (see, e.g., [94,120]), but we restrict attention in these notes to (1.2) for the sake of simplicity.

The Self-Avoiding Walk

The self-avoiding walk is a model of fundamental interest in combinatorics, probability theory, statistical physics and polymer chemistry. It is a model of random walk paths but it cannot be described in terms of transition probabilities and thus is not even a stochastic process. It is certainly non-Markovian. These features makes the subject difficult, and many of the central problems remain unsolved. See [127, 158] for extensive surveys.

The self-avoiding walk is a basic example in the theory of critical phenomena, due to its close links with models of ferromagnetism such as the Ising model. In particular, it can be understood as the $N \to 0$ limit of the N-vector model [79] (see also [158, Sect. 2.3]). In polymer chemistry, self-avoiding walks are used to model a single linear polymer molecule in a good solution [80, 205]. The flexibility of the polymer is modelled by the possible configurations of a self-avoiding walk, while the self-avoidance constraint models the excluded volume effect that causes the polymer to repel itself.

In this chapter, we first give an overview of the self-avoiding walk and its predicted asymptotic behaviour. Then we define the bubble condition and show that it is a sufficient condition for a particular critical exponent (namely γ) to exist and take its mean-field value.

2.1 Asymptotic Behaviour

An n-step self-avoiding walk starting at x and ending at y is an n-step walk $(\omega(0), \omega(1), \ldots, \omega(n))$ with $\omega(0) = x$, $\omega(n) = y$, and $\omega(i) \neq \omega(j)$ for all $i \neq j$. We will assume for simplicity that the walks take steps in Ω given either by (1.1) or (1.2). Let $\mathcal{S}_n(x, y)$ be the set of n-step self-avoiding walks from x to y, let $\mathcal{S}_n = \cup_{x \in \mathbb{Z}^d} \mathcal{S}_n(0, x)$ denote the set of all n-step self-avoiding walks starting from the origin, and let $\mathcal{S}(x, y) = \cup_{n=0}^{\infty} \mathcal{S}_n(x, y)$ denote the set of all self-avoiding walks of any length from x to y. Let $c_n(x, y)$ denote the cardinality of $\mathcal{S}_n(x, y)$. In particular, $c_0(x, y) = \delta_{x,y}$. We will use the abbreviations $\mathcal{S}_n(x) = \mathcal{S}_n(0, x)$, $c_n(x) = c_n(0, x)$, and $c_n = \sum_{x \in \mathbb{Z}^d} c_n(x)$.

Thus c_n counts the number of n-step self-avoiding walks that start at the origin and end anywhere.

More generally, given a walk ω, let

$$U_{st}(\omega) = \begin{cases} -1 \text{ if } \omega(s) = \omega(t) \\ 0 \text{ if } \omega(s) \neq \omega(t), \end{cases} \tag{2.1}$$

and, for $\lambda \in [0, 1]$, let

$$c_n^{(\lambda)}(x) = \sum_{\omega \in \mathcal{W}_n(x)} \prod_{0 \leq s < t \leq n} (1 + \lambda U_{st}(\omega)). \tag{2.2}$$

For $\lambda = 0$, (2.2) is the same as the quantity $c_n^{(0)}(x)$ defined previously. For $\lambda = 1$, we have $c_n^{(1)}(x) = c_n(x)$, and we will usually omit the superscript $^{(1)}$ when $\lambda = 1$. For $0 < \lambda < 1$, (2.2) defines a much-studied model of weakly self-avoiding walks (sometimes called the Domb–Joyce model after [64]) in which walks with self intersections receive less weight than walks that are self-avoiding.

For $\lambda \in [0, 1]$, let

$$c_n^{(\lambda)} = \sum_{x \in \mathbb{Z}^d} c_n^{(\lambda)}(x). \tag{2.3}$$

Since $1 + \lambda U_{st}(\omega) \leq 1$ for all s, t, ω, we have

$$\prod_{0 \leq s < t \leq m+n} (1 + \lambda U_{st}(\omega)) \leq \prod_{0 \leq s < t \leq m} (1 + \lambda U_{st}(\omega)) \prod_{m \leq s < t \leq m+n} (1 + \lambda U_{st}(\omega)), \tag{2.4}$$

from which we easily conclude that

$$c_{m+n}^{(\lambda)} \leq c_m^{(\lambda)} c_n^{(\lambda)}. \tag{2.5}$$

Therefore, $\log c_n^{(\lambda)}$ is a subadditive sequence. By a standard lemma [158, Lemma 1.2.2], it follows that the limit

$$\mu_\lambda = \lim_{n \to \infty} \left(c_n^{(\lambda)} \right)^{1/n} \tag{2.6}$$

exists and that moreover

$$c_n^{(\lambda)} \geq \mu_\lambda^n. \tag{2.7}$$

When $\lambda = 1$, $\mu = \mu_1$ is known as the *connective constant*. For nearest-neighbour walks, it is easy to see that $d \leq \mu \leq 2d - 1$. The lower bound follows from the fact that c_n is at least as large as the number d^n of walks that take steps only in the positive coordinate directions. The upper bound follows from the fact that c_n is at most the number $2d(2d - 1)^{n-1}$ of walks that never reverse a previous step. The exact value of μ is not known in general, although good rigorous numerical upper and lower bounds have been obtained [55, 106, 171, 187]. Numerical estimates of μ are 2.638 and 4.684 for nearest-neighbour self-avoiding walks in dimensions 2 and 3 respectively—in fact there are higher precision estimates due to A.J. Guttmann and coworkers. It has been conjectured [168–170], and been confirmed by numerical evidence [70], that on the 2-dimensional hexagonal lattice $\mu = \sqrt{2 + \sqrt{2}}$. It has been observed from enumeration data that on the 2-dimensional square lattice μ is very well approximated by the reciprocal of the smallest positive root of the quartic equation $581x^4 + 7x^2 - 13 = 0$ [54, 137], although no derivation or explanation of this equation has been discovered.

For the nearest-neighbour self-avoiding walk on $\mathbb{Z}^d$, the lace expansion has been used to prove that $\mu(d)$ has an asymptotic expansion to all orders in $1/d$, with integer coefficients, and that

$$\mu(d) = 2d - 1 - \frac{1}{2d} - \frac{3}{(2d)^2} - \frac{16}{(2d)^3} - \frac{102}{(2d)^4} - \frac{729}{(2d)^5}$$
$$- \frac{5533}{(2d)^6} - \frac{42229}{(2d)^7} - \frac{288761}{(2d)^8} + O\left(\frac{1}{(2d)^9}\right). \qquad (2.8)$$

Without using the lace expansion (which was not yet invented), the above coefficients were computed in [74], up to and including $-102(2d)^{-4}$, without a rigorous estimate for the error. About the same time, the formula $\mu(d) = 2d - 1 - (2d)^{-1} + O((2d)^{-2})$ was proved in [140]. In [101, 102], the lace expansion was used to prove the existence of an asymptotic expansion to all orders, and also that $\mu(d) = 2d - 1 - (2d)^{-1} - 3(2d)^{-2} - 16(2d)^{-3} - 102(2d)^{-4} + O\left((2d)^{-5}\right)$. The four additional coefficients in (2.8) were obtained in [53]. It seems likely that the asymptotic expansion has radius of convergence zero, though there is no proof of this. For further $1/d$ expansion results (but without rigorous error estimates) see [77, 163, 164]. For asymptotics of the connective constant for the spread-out model, as $L \to \infty$, see [118, 176].

For $\lambda = 0$, we have seen in Chap. 1 that $c_n^{(0)} = |\Omega|^n$, and thus the number of n-step walks grows purely exponentially in n. There is overwhelming evidence to support the belief that for $\lambda \in (0, 1]$, the asymptotic form of $c_n^{(\lambda)}$ is given by

$$c_n^{(\lambda)} \sim A_\lambda \mu_\lambda^n n^{\gamma - 1}. \qquad (2.9)$$

Here, A_λ is a constant which, like μ_λ, depends on λ, d and Ω, but the critical exponent γ is independent of λ and Ω and is given by

$$\gamma = \begin{cases} 1 & \text{if } d = 1 \\ \frac{43}{32} & \text{if } d = 2 \\ 1.162... & \text{if } d = 3 \\ 1 \text{ with logarithmic corrections if } d = 4 \\ 1 & \text{if } d \geq 5. \end{cases} \qquad (2.10)$$

The conjectured logarithmic correction in four dimensions, predicted by the renormalization group method, is

$$c_n^{(\lambda)} \sim A_\lambda \mu_\lambda^n (\log n)^{1/4} \quad \text{if } d = 4. \tag{2.11}$$

The independence of γ on $\lambda \in (0,1]$ and Ω is referred to as universality. Similarly, the power of the logarithm in (2.11) is believed to be universal.

The exponent γ has the following probabilistic interpretation. Consider the case $\lambda = 1$, and let q_n denote the probability that two independent n-step self-avoiding walks started at the origin do not have any intersection apart from their common starting point. Since a non-intersecting pair of n-step self-avoiding walks comprises a single $2n$-step self-avoiding walk, if (2.9) holds then

$$q_n = \frac{c_{2n}}{c_n^2} \sim \frac{2^{\gamma-1}}{A_1} \frac{1}{n^{\gamma-1}}. \tag{2.12}$$

In dimensions $d > 4$, the lace expansion has been used to prove that (2.9) holds with $\gamma = 1$ for various choices of λ and Ω, including the nearest-neighbour model with $\lambda = 1$ [96–98]. Note that $\gamma = 1$ corresponds to purely exponential growth on the right hand side of (2.9), as is the case for the simple random walk. Also, there is no decay as $n \to \infty$ in (2.12) when $\gamma = 1$. Partial results for the 4-dimensional case have been obtained in [43, 44, 129] (physics references include [38, 65]). The 3-dimensional case is completely unsolved mathematically. Evidence strongly supporting the value $\gamma = \frac{43}{32}$, which was first predicted in [168–170], has been obtained in [150], by associating the 2-dimensional self-avoiding walk with $\text{SLE}_{8/3}$. Numerical tests supporting the role of $\text{SLE}_{8/3}$ in the description of the 2-dimensional self-avoiding walk can be found in [139]. For $d = 1$, the strictly self-avoiding nearest-neighbour model is trivial and $c_n^{(1)} = 2$ for all $n \geq 1$, so $\gamma = 1$. For the 1-dimensional strictly self-avoiding spread-out model, or for the weakly self-avoiding walk for any $\lambda \in (0,1)$, the determination of $c_n(\lambda)$ is no longer trivial, but has been analyzed in detail (see [16, 84, 116, 146]).

For $d = 2, 3, 4$, the best upper bounds on c_n (with $\lambda = 1$) are still the forty-year-old bounds

$$\mu^n \leq c_n \leq \begin{cases} \mu^n \exp[Kn^{1/2}] & \text{if } d = 2 \\ \mu^n \exp[Kn^{2/(2+d)} \log n] & \text{if } d = 3, 4 \end{cases} \tag{2.13}$$

for a positive constant K [88, 140] (see also [158, Chapter 3]). This is a long way from (2.9).

We can define a measure on $\mathcal{W}_n$ by

$$\mathbb{E}_n^{(\lambda)} X = \frac{1}{c_n^{(\lambda)}} \sum_{\omega \in \mathcal{W}_n} X(\omega) \prod_{0 \leq s < t \leq n} (1 + \lambda U_{st}(\omega)). \tag{2.14}$$

Exercise 2.1. For $\lambda = 1$, the above measure is the uniform measure on $\mathcal{S}_n$. A family of probability measures P_m on $\mathcal{S}_n$ is called *consistent* if $P_m(\omega) =$

$\sum_{\rho > \omega} P_n(\rho)$ for all $n \geq m$ and for all $\omega \in \mathcal{S}_n$, where the sum is over all ρ whose first m steps agree with ω. Show that the uniform measure does not provide a consistent family.

The mean-square displacement is $\mathbb{E}_n^{(\lambda)} |\omega(n)|^2$ and it is believed that

$$\mathbb{E}_n^{(\lambda)} |\omega(n)|^2 \sim v_\lambda n^{2\nu} \tag{2.15}$$

where v_λ is a constant depending on λ, d, Ω, and where ν is universal and given by

$$\nu = \begin{cases} 1 & \text{if } d = 1 \\ \frac{3}{4} & \text{if } d = 2 \\ 0.588... & \text{if } d = 3 \\ \frac{1}{2} \text{ with logarithmic corrections} & \text{if } d = 4 \\ \frac{1}{2} & \text{if } d \geq 5. \end{cases} \tag{2.16}$$

The conjectured logarithmic correction to ν in four dimensions, predicted by the renormalization group, is

$$\mathbb{E}_n^{(\lambda)} |\omega(n)|^2 \sim v_\lambda n (\log n)^{1/4} \quad \text{if } d = 4. \tag{2.17}$$

In dimensions $d > 4$, the lace expansion has been used to prove that (2.15) holds with $\nu = 1/2$ for various choices of λ and Ω, including the nearest-neighbour model with $\lambda = 1$ [97, 98]. Partial results for $d = 4$ have been obtained in [43, 44, 129]. For $d = 2, 3, 4$, for the nearest-neighbour model with $\lambda = 1$, it is still an open problem even to prove the "obvious" bounds that the mean-square displacement is bounded below by n (cf. (1.16)) or bounded above by const $n^{2-\epsilon}$ for some $\epsilon > 0$. For $d = 1$, the ballistic behaviour $\nu = 1$ is obvious for the strictly self-avoiding nearest-neighbour model. It is not obvious that $\nu = 1$ for the 1-dimensional strictly self-avoiding spread-out model, or for the 1-dimensional weakly self-avoiding walk, but ballistic behaviour has been proved also in these cases [84, 146].

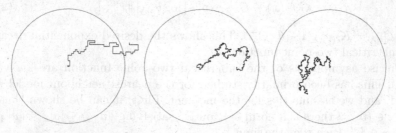

Fig. 2.1. Nearest-neighbour self-avoiding walks on $\mathbb{Z}^2$ taking $n = 100$, 1,000 and 10,000 steps, generated using the pivot algorithm [159]. The circles have radius $n^{3/4}$, in units of the step size of the self-avoiding walk.

The *two-point function* is defined by

$$G_z^{(\lambda)}(x) = \sum_{n=0}^{\infty} c_n^{(\lambda)}(x) z^n \qquad (2.18)$$

and the *susceptibility* by

$$\chi^{(\lambda)}(z) = \sum_{x \in \mathbb{Z}^d} G_z^{(\lambda)}(x) = \sum_{n=0}^{\infty} c_n^{(\lambda)} z^n. \qquad (2.19)$$

The latter has radius of convergence $z_c^{(\lambda)} = 1/\mu_\lambda$, by (2.6). For $\lambda = 1$, a proof that the two-point function also has this radius of convergence is given in [158, Corollary 3.2.6].

Exercise 2.2. Show that the 1-dimensional strictly self-avoiding walk ($\lambda = 1$) two-point function is given by

$$\hat{G}_z(k) = \frac{1 - z^2}{1 + z^2 - 2z \cos k}. \qquad (2.20)$$

For $\lambda \in [0, 1]$ and $z \in (0, z_c^{(\lambda)})$, the two-point function decays exponentially. To see this for the nearest-neighbour model, we note that $c_n^{(\lambda)}(x) = 0$ for $n < \|x\|_\infty$, and hence

$$G_z^{(\lambda)}(x) = \sum_{n=\|x\|_\infty}^{\infty} c_n^{(\lambda)}(x) z^n \leq \sum_{n=\|x\|_\infty}^{\infty} c_n^{(\lambda)} z^n. \qquad (2.21)$$

Since $\left(c_n^{(\lambda)}\right)^{1/n} \to \mu_\lambda$ by (2.6), for any $\epsilon > 0$ there is a positive $K_{\epsilon,\lambda}$ such that

$$c_n^{(\lambda)} \leq K_{\epsilon,\lambda}(\mu_\lambda + \epsilon)^n \qquad (2.22)$$

for all $n \geq 1$. Given a positive $z < z_c^{(\lambda)} = \mu_\lambda^{-1}$, we choose $\epsilon(z) > 0$ such that $\theta_{z,\lambda} = (\mu_\lambda + \epsilon(z))z < 1$. Then substitution of (2.22) into (2.21) gives

$$G_z^{(\lambda)}(x) \leq C_{z,\lambda} \exp[-|\log \theta_{z,\lambda}| \, \|x\|_\infty], \qquad (2.23)$$

with $C_{z,\lambda} = K_{\epsilon(z),\lambda}(1 - \theta_{z,\lambda})^{-1}$. This shows the desired exponential decay of the subcritical two-point function.

Precise asymptotics of the subcritical two-point function are known in detail. This has been primarily studied for the nearest-neighbour model with $\lambda = 1$, and we assume this for the moment. First, it can be shown that for each $z \in (0, z_c)$ there is a norm $|\cdot|_z$ on $\mathbb{R}^d$ satisfying $\|u\|_\infty \leq |u|_z \leq \|u\|_1$ for every $u \in \mathbb{R}^d$, such that the limit

$$m(z) = \lim_{|x|_z \to \infty} \frac{-\log G_z(x)}{|x|_z} \qquad (2.24)$$

exists and is finite [158, Theorem 4.1.18]. The *correlation length* is defined by

$$\xi(z) = \frac{1}{m(z)}. \tag{2.25}$$

Detailed asymptotics of the subcritical two-point function, known as Ornstein–Zernike decay, were obtained in [48,130]. It is known that $\xi(z) \to \infty$ as $z \to z_c^-$ (see, e.g., [158, Corollary 4.1.15]), and it is predicted that

$$\xi(z) \sim \text{const} \frac{1}{(1 - z/z_c)^\nu} \quad \text{as } z \to z_c^-, \tag{2.26}$$

with the same exponent ν as in (2.16).

For $\lambda \in (0,1]$, it is predicted that the exponential decay of the subcritical two-point function is replaced at $z = z_c$ by

$$G_{z_c}^{(\lambda)}(x) \sim \text{const} \frac{1}{|x|^{d-2+\eta}} \quad \text{as } |x| \to \infty \tag{2.27}$$

and

$$\hat{G}_{z_c}^{(\lambda)}(k) \sim \text{const} \frac{1}{|k|^{2-\eta}} \quad \text{as } k \to 0 \tag{2.28}$$

with η given in terms of γ and ν by Fisher's relation $\gamma = (2 - \eta)\nu$ (and with no logarithmic correction for $d = 4$, to leading order). Equation (2.27) has been proved (with $\eta = 0$) for the nearest-neighbour model in dimensions $d \geq 5$ [90], using the lace expansion. The k-space asymptotics are easier and are also known for the nearest-neighbour model when $d \geq 5$. Equation (2.27) has also been proved for the spread-out model with $d > 4$ and L sufficiently large [91]. In [39,44], (2.27) is proved for a 4-dimensional hierarchical model with λ sufficiently small (again with $\eta = 0$).

It is also believed that, for all $\lambda \in (0,1]$,

$$\chi^{(\lambda)}(z) \sim \text{const} \frac{1}{(1 - z/z_c)^\gamma} \quad \text{as } z \to z_c^-, \tag{2.29}$$

with a multiplicative factor $[\log(1 - z/z_c)]^{1/4}$ when $d = 4$. This has been proved using the lace expansion for the nearest-neighbour model with $d > 4$ and $\lambda = 1$, with $\gamma = 1$ [97,98]. In Sect. 5.4, we will see how to prove (2.29), with $\gamma = 1$, for the spread-out model with $d > 4$, $\lambda = 1$, and L sufficiently large, and for the nearest-neighbour model with $\lambda = 1$ and d sufficiently large.

The scaling limit, assuming it exists, is the law of the path $n^{-\nu}\omega$ in the limit $n \to \infty$ (a factor $(\log n)^{-1/4}$ should be included for $d = 4$), where ω is an n-step self-avoiding walk. The scaling limit is believed not to depend on $\lambda \in (0,1]$ in any important way. This limit is conjectured to be $\text{SLE}_{8/3}$ for $d = 2$, and the limit is not understood for $d = 3$. For $d = 4$, the scaling limit is believed to be Brownian motion, and for $d \geq 5$, the lace expansion has been used to prove that the scaling limit is Brownian motion [97,98].

The special role of $d = 4$ for the asymptotics of the self-avoiding walk is summarized by saying that $d = 4$ is the *upper critical dimension*, and that *mean-field* behaviour applies when $d > 4$. Above $d = 4$, the self-avoiding walk has the same leading asymptotics as the simple random walk. Logarithmic corrections to simple random walk behaviour occur when $d = 4$, and different power laws appear for $d < 4$.

The critical nature of $d = 4$ can be guessed from the fact that Brownian motion is 2-dimensional. Since two 2-dimensional sets generically do not intersect in more than $4 = 2 + 2$ dimensions, above four dimensions the self-avoidance constraint does not play an important role.

2.2 Differential Inequalities and the Bubble Condition

We now define the bubble condition and show that it is a sufficient condition for a particular critical exponent (namely γ) to exist and take its mean-field value. This is a useful precursor to the lace expansion. It is also a useful precursor to the study of lattice trees and percolation, where the bubble condition will be replaced by the square and triangle conditions, respectively.

For simplicity, we restrict attention in this section to the strictly self-avoiding walk with $\lambda = 1$. We fix Ω to be either (1.1) or (1.2).

The *bubble diagram* is defined by

$$B(z) = \sum_{x \in \mathbb{Z}^d} G_z(x)^2. \qquad (2.30)$$

The name "bubble diagram" comes from a Feynman diagram notation in which the two-point function evaluated at vertices x and y is denoted by a line terminating at x and y. In this notation,

$$B(z) = \sum_{x \in \mathbb{Z}^d} 0 \, \bigcirc \, x = 0 \bigcirc \,,$$

where in the diagram on the right it is implicit that the unlabelled vertex is summed over $\mathbb{Z}^d$. The bubble diagram can be rewritten in terms of the Fourier transform of the two-point function, using (2.30) and the Parseval relation, as

$$B(z) = \|G_z\|_2^2 = \|\hat{G}_z\|_2^2 = \int_{[-\pi,\pi]^d} \hat{G}_z(k)^2 \, \frac{\mathrm{d}^d k}{(2\pi)^d}. \qquad (2.31)$$

The *bubble condition* is the statement that $B(z_c) < \infty$. In other words, the bubble condition states that $\hat{G}_{z_c}(k)$ is square integrable. Recall that square integrability of $\hat{C}_{1/|\Omega|}(k)$ was important in Exercise 1.7.

In view of the definition of η in (2.27) or (2.28), it follows from (2.31) that the bubble condition is satisfied provided $\eta > (4 - d)/2$. Hence the bubble condition for $d > 4$ is implied by the *infrared bound* $\eta \geq 0$. If the values for η arising from Fisher's relation and the conjectured values of γ and ν are correct, then the bubble condition will not hold in dimensions $2, 3$ or 4, with the divergence of the bubble diagram being only logarithmic in four dimensions.

Throughout these notes,

$$f(z) \simeq g(z) \quad \text{denotes} \quad c^{-1}g(z) \le f(z) \le cg(z) \tag{2.32}$$

for some $c > 0$, uniformly in $z < z_c$. In this section, we prove a differential inequality for the susceptibility, which shows that the bubble condition implies that $\gamma = 1$ in the sense that $\chi(z) \simeq (1 - z/z_c)^{-1}$. In fact, the lower bound

$$\chi(z) \ge \frac{z_c}{z_c - z} \tag{2.33}$$

is an immediate consequence of (2.19) and the subadditivity bound $c_n \ge \mu^n = z_c^{-n}$, and holds with or without the bubble condition. It remains to prove that the complementary upper bound

$$\chi(z) \le \text{const} \frac{1}{z_c - z} \tag{2.34}$$

is a consequence of the bubble condition. This will be shown in the following theorem. In Chap. 5, we will use the lace expansion to prove the bubble condition for the spread-out model (1.2) for $d > 4$ with L sufficiently large, and for the nearest-neighbour model with d sufficiently large.

A version of Theorem 2.3 was proved in [36]. The role of the bubble condition in proving mean-field behaviour for spin systems in dimensions $d > 4$ was developed previously, in [3, 76, 192] (see also [73]). In Sect. 5.4, it will be shown that the lace expansion actually provides a differential *equality* in place of the differential inequalities of Theorem 2.3.

Theorem 2.3. *For $0 < z < z_c$, the susceptibility obeys the differential inequalities*

$$\frac{\chi(z)^2}{B(z)} \le \frac{\mathrm{d}}{\mathrm{d}z}[z\chi(z)] \le \chi(z)^2 \tag{2.35}$$

and the inequalities

$$\frac{z_c}{z_c - z} \le \chi(z) \le B(z_c)\frac{2z_c - z}{z_c - z}. \tag{2.36}$$

Thus the bubble condition implies that $\chi(z) \simeq (1 - z/z_c)^{-1}$, which is to say that γ exists and equals 1.

Proof. We first prove the differential inequalities (2.35). By definition,

$$\chi(z) = \sum_{n=0}^{\infty} c_n z^n = \sum_{y} \sum_{\omega \in \mathcal{S}(0,y)} z^{|\omega|}, \tag{2.37}$$

where $|\omega|$ denotes the number of steps in ω. For $0 < z < z_c$, term by term differentiation gives

$$Q(z) = \frac{\mathrm{d}}{\mathrm{d}z}[z\chi(z)] = \sum_{y} \sum_{\omega \in \mathcal{S}(0,y)} (|\omega| + 1)z^{|\omega|}, \qquad (2.38)$$

where the first equality defines $Q(z)$. This can be rewritten as

$$Q(z) = \sum_{y} \sum_{\omega \in \mathcal{S}(0,y)} \sum_{x} I[\omega(j) = x \text{ for some } j]z^{|\omega|}$$

$$= \sum_{x,y} \sum_{\substack{\omega^{(1)} \in \mathcal{S}(0,x) \\ \omega^{(2)} \in \mathcal{S}(x,y)}} z^{|\omega^{(1)}| + |\omega^{(2)}|} I[\omega^{(1)} \cap \omega^{(2)} = \{x\}], \qquad (2.39)$$

where I denotes the indicator function.

If we ignore the mutual avoidance of $\omega^{(1)}$ and $\omega^{(2)}$ in (2.39), we obtain the upper bound

$$\frac{\mathrm{d}(z\chi(z))}{\mathrm{d}z} \le \chi(z)^2 \qquad (2.40)$$

of (2.35).

To obtain a complementary bound, we rewrite $Q(z)$ by using the inclusion-exclusion relation in the form

$$I[\omega^{(1)} \cap \omega^{(2)} = \{x\}] = 1 - I[\omega^{(1)} \cap \omega^{(2)} \ne \{x\}].$$

This gives

$$Q(z) = \chi(z)^2 - \sum_{x,y} \sum_{\substack{\omega^{(1)} \in \mathcal{S}(0,x) \\ \omega^{(2)} \in \mathcal{S}(x,y)}} z^{|\omega^{(1)}| + |\omega^{(2)}|} I[\omega^{(1)} \cap \omega^{(2)} \ne \{x\}]. \qquad (2.41)$$

Since $x \in \omega^{(1)} \cap \omega^{(2)}$, the indicator forces a nontrivial intersection. In the last term on the right hand side of (2.41), let $w = \omega^{(2)}(l)$ be the site of the last intersection of $\omega^{(2)}$ with $\omega^{(1)}$, where time is measured along $\omega^{(2)}$ beginning at its starting point x. Then the portion of $\omega^{(2)}$ corresponding to times greater than l must avoid all of $\omega^{(1)}$. Relaxing the restrictions that this portion of $\omega^{(2)}$ avoid both the remainder of $\omega^{(2)}$ and the part of $\omega^{(1)}$ linking w to x, and also relaxing the mutual avoidance of the two portions of $\omega^{(1)}$, gives the upper bound

$$\sum_{x,y} \sum_{\substack{\omega^{(1)} \in \mathcal{S}(0,x) \\ \omega^{(2)} \in \mathcal{S}(x,y)}} z^{|\omega^{(1)}| + |\omega^{(2)}|} I[\omega^{(1)} \cap \omega^{(2)} \ne \{x\}] \le Q(z)[B(z) - 1], \qquad (2.42)$$

as illustrated in Fig. 2.2. Here the factor $B(z) - 1$ arises from the two paths joining w and x. The upper bound involves $B(z) - 1$ rather than $B(z)$ since there will be no contribution from the $x = 0$ term in (2.30).

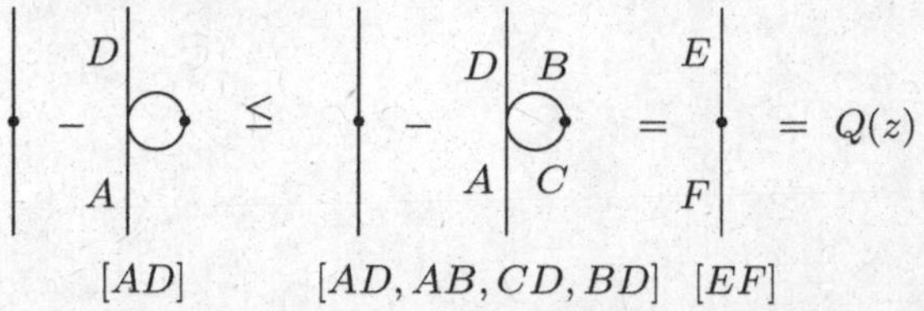

$$\left|\begin{matrix} D \\ \bigcirc \\ A \end{matrix}\right| \leq \left|\begin{matrix} D\,|\,B \\ \bigcirc \\ A\,|\,C \end{matrix}\right| = \left|\begin{matrix} E \\ \\ F \end{matrix}\right| = Q(z)$$

$$[AD] \qquad [AD, AB, CD, BD] \quad [EF]$$

Fig. 2.2. A diagrammatic representation of the inequality $\chi(\dot z)^2 - Q(z)[B(z) - 1] \leq Q(z)$. The lists of pairs of lines indicate interactions, in the sense that the corresponding walks must avoid each other.

Exercise 2.4. Convince yourself that (2.42) is correct.

Combining (2.41) and (2.42) gives

$$Q(z) \geq \chi(z)^2 - Q(z)[B(z) - 1]. \tag{2.43}$$

Solving for $Q(z)$ gives

$$Q(z) \geq \frac{\chi(z)^2}{B(z)}, \tag{2.44}$$

which is the lower bound of (2.35).

Next, we show that (2.35) implies (2.36). The lower bound of (2.36) has already been established in (2.33) (and also follows by integration of the upper bound of (2.35)). To obtain the upper bound of (2.36) from the lower bound of (2.35), we proceed as follows. Let $z_1 \in [0, z_c)$. The lower bound of (2.35) implies that, for $z \in [z_1, z_c)$,

$$z\left(-\frac{d\chi^{-1}}{dz}\right) \geq \frac{1}{B(z)} - \frac{1}{\chi(z)} \geq \frac{1}{B(z_c)} - \frac{1}{\chi(z_1)}, \tag{2.45}$$

where χ^{-1} denotes the reciprocal. We bound the factor of z on the left hand side by z_c and then integrate from z_1 to z_c. Using the fact that $\chi(z_c)^{-1} = 0$ by (2.33), this gives

$$z_c\chi(z_1)^{-1} \geq [B(z_c)^{-1} - \chi(z_1)^{-1}](z_c - z_1). \tag{2.46}$$

Rewriting gives the upper bound of (2.36). ∎

By (2.39), $Q(z)$ is the generating function for pairs of self-avoiding walks which do not intersect each other apart from their common starting point. It follows from Theorem 2.3 that if the bubble condition holds then

$$\frac{Q(z)}{\chi(z)^2} \simeq 1, \tag{2.47}$$

a relation related to the non-vanishing of the non-intersection probability q_n of (2.12), as $n \to \infty$, when $\gamma = 1$.

3

The Lace Expansion for the Self-Avoiding Walk

The lace expansion was derived by Brydges and Spencer in [45]. Their derivation, which is given below in Sects. 3.2–3.3, involves an expansion and resummation procedure closely related to the cluster expansions of statistical mechanics [40]. It was later noted that the lace expansion can also be seen as resulting from repeated application of the inclusion-exclusion relation [186]. For a more combinatorial description of the lace expansion, see [211]. We first discuss the inclusion-exclusion approach.

3.1 Inclusion-Exclusion

The inclusion-exclusion approach to the lace expansion is closely related to the method of proof of Theorem 2.3. In that proof, a single inclusion-exclusion was used to obtain upper and lower bounds. Here, we will derive an identity by using repeated inclusion-exclusion.

For simplicity, we restrict attention to the strictly self-avoiding walk ($\lambda = 1$). We consider a walk taking steps in a finite set Ω, so that $\omega(i+1) - \omega(i) \in \Omega$ for each i, but there is no need here for a symmetry assumption and Ω is an arbitrary finite set. As in (1.10), we write

$$D(x) = \frac{1}{|\Omega|} I[x \in \Omega]. \tag{3.1}$$

We rewrite $c_n(x)$ using the inclusion-exclusion relation. Namely, we first count all walks from 0 to x which are self-avoiding *after* the first step, and then subtract the contribution due to those which are not self-avoiding from the beginning, i.e., walks that return to the origin. Since $c_1(0, y) = 1$ for $y \in \Omega$, this gives

$$c_n(x) = (c_1 * c_{n-1})(x) - \sum_{y \in \Omega} \sum_{\omega^{(1)} \in \mathcal{S}_{n-1}(y,x)} I[0 \in \omega^{(1)}]. \tag{3.2}$$

Comparing with (1.5), it is the second term on the right hand side that makes the above equation interesting.

The inclusion-exclusion relation can now be applied to the last term of (3.2), as follows. Let s be the first (and only) time that $\omega^{(1)}(s) = 0$. Then for $y \in \Omega$,

$$\sum_{\omega^{(1)} \in \mathcal{S}_{n-1}(y,x)} I[0 \in \omega^{(1)}]$$

$$= \sum_{s=1}^{n-1} \sum_{\substack{\omega^{(2)} \in \mathcal{S}_s(y,0) \\ \omega^{(3)} \in \mathcal{S}_{n-1-s}(0,x)}} I[\omega^{(2)} \cap \omega^{(3)} = \{0\}] \tag{3.3}$$

$$= \sum_{s=1}^{n-1} \left[c_s(y,0) c_{n-1-s}(0,x) - \sum_{\substack{\omega^{(2)} \in \mathcal{S}_s(y,0) \\ \omega^{(3)} \in \mathcal{S}_{n-1-s}(0,x)}} I[\omega^{(2)} \cap \omega^{(3)} \neq \{0\}] \right].$$

We can interpret $c_s(y,0)$ as the number of $(s+1)$-step walks which step from the origin directly to y, then return to the origin in s steps, and which have distinct vertices apart from the fact that they return to their starting point. Let $\mathcal{U}_s$ denote the set of all s-step self-avoiding loops at the origin (s-step walks which begin and end at the origin but which otherwise have distinct vertices), and let u_s be the cardinality of $\mathcal{U}_s$. Then

$$\sum_{y \in \Omega} \sum_{\omega^{(1)} \in \mathcal{S}_{n-1}(y,x)} I[0 \in \omega^{(1)}]$$

$$= \sum_{s=2}^{n} u_s c_{n-s}(x) - \sum_{s=2}^{n} \sum_{\substack{\omega^{(2)} \in \mathcal{U}_s \\ \omega^{(3)} \in \mathcal{S}_{n-s}(0,x)}} I[\omega^{(2)} \cap \omega^{(3)} \neq \{0\}]. \tag{3.4}$$

Continuing in this fashion, in the last term on the right hand side of the above equation, let $t \geq 1$ be the first time along $\omega^{(3)}$ that $\omega^{(3)}(t) \in \omega^{(2)}$, and let $v = \omega^{(3)}(t)$. Then the inclusion-exclusion relation can be applied again to remove the avoidance between the portions of $\omega^{(3)}$ before and after t, and correct for this removal by the subtraction of a term involving a further intersection. Repetition of this procedure leads to the convolution equation

$$c_n(0,x) = (|\Omega| D * c_{n-1})(x) + \sum_{m=2}^{n} (\pi_m * c_{n-m})(x), \tag{3.5}$$

where we have used $c_1(x) = |\Omega| D(x)$, and where π_m is given by

$$\pi_m(v) = \sum_{N=1}^{\infty} (-1)^N \pi_m^{(N)}(v), \tag{3.6}$$

with the terms on the right hand side defined as follows. The $N = 1$ term is given by

$$\pi_m^{(1)}(v) = \delta_{0,v} u_m = \delta_{0,v} \ \ 0 \bigcirc ,$$

where the diagram represents u_m. The $N = 2$ term is

$$\pi_m^{(2)}(v) = \sum_{\substack{m_1, m_2, m_3 : \\ m_1 + m_2 + m_3 = m}} \sum_{\omega_1 \in \mathcal{S}_{m_1}(0,v)} \sum_{\omega_2 \in \mathcal{S}_{m_2}(v,0)} \sum_{\omega_3 \in \mathcal{S}_{m_3}(0,v)} I(\omega_1, \omega_2, \omega_3),$$

where $I(\omega_1, \omega_2, \omega_3)$ is equal to 1 if the ω_i are pairwise mutually avoiding apart from their common endpoints, and otherwise equals 0. Diagrammatically this can be represented by

$$\pi_m^{(2)}(v) = 0 \bigoplus v ,$$

where each line represents a sum over self-avoiding walks between the endpoints of the line, with mutual avoidance between the three pairs of lines in the diagram. Similarly

$$\pi_m^{(3)}(v) = \underset{0 \qquad v}{\bigotimes} ,$$

where now there is mutual avoidance between some but not all pairs of lines in the diagram; a precise description requires some care. The unlabelled vertex is summed over $\mathbb{Z}^d$. A slashed diagram line is used to indicate a walk which may have zero steps, i.e., be a single site, whereas lines without a slash correspond to walks of at least one step. All the higher order terms can be expressed as diagrams in this way, and with some care it is possible to keep track of the pattern of mutual avoidance between subwalks (individual lines in the diagram) which emerges. The algebraic derivation of the expansion, described next, keeps track of this mutual avoidance automatically. Equations (3.5)–(3.6) constitute the lace expansion. No laces have appeared yet, but they will come later.

Exercise 3.1. Determine a precise expression for $\pi_m^{(3)}(v)$. What is the picture for $\pi_m^{(4)}(v)$?

3.2 Expansion

In this and the following section, we give the original derivation of the lace expansion due to Brydges and Spencer [45]. The expansion applies in a more general context than we have considered so far, and we will give a quite general derivation.

Consider walks taking steps in a finite subset $\Omega \subset \mathbb{Z}^d$. Suppose that to each walk $\omega = (\omega(0), \omega(1), \ldots, \omega(n))$ and each pair $s, t \in \{0, 1, \ldots, n\}$, we are given a complex number $\mathcal{U}_{st}(\omega)$ (for example, (2.1)).

Definition 3.2. (i) *Given an interval $I = [a, b]$ of positive integers, we refer to a pair $\{s, t\}$ $(s < t)$ of elements of I as an* edge. *To abbreviate the notation, we usually write st for $\{s, t\}$. A set of edges is called a* graph. *The set of all graphs on $[a, b]$ is denoted $\mathcal{B}[a, b]$.*
(ii) *A graph Γ is said to be* connected *if both a and b are endpoints of edges in Γ, and if in addition, for any $c \in (a, b)$, there are $s, t \in [a, b]$ such that $s < c < t$ and $st \in \Gamma$. In other words, Γ is connected if, as intervals, $\cup_{st \in \Gamma}(s, t) = (a, b)$. The set of all connected graphs on $[a, b]$ is denoted $\mathcal{G}[a, b]$.*

An apology is required for graph theorists. The above notion of connectivity is not the usual notion of path-connectivity in graph theory. Instead, the above notion relies heavily on the fact that the vertices of the graph are linearly ordered in time, and may be justified by the fact that connected graphs are those for which $\cup_{st \in \Gamma}(s, t)$ is equal to the connected interval (a, b). In any event, it is decidedly not path-connectivity. There are connected graphs that are not path-connected, and vice versa. It is convenient to have in mind the representation of graphs illustrated in Fig. 3.1.
We set $K[a, a] = 1$, and for $a < b$ we define

$$K[a, b] = \prod_{a \le s < t \le b} (1 + \mathcal{U}_{st}),$$

(3.7)

where the dependence on ω is left implicit. By expanding the product in (3.7), we obtain

$$K[a, b] = \sum_{\Gamma \in \mathcal{B}[a, b]} \prod_{st \in \Gamma} \mathcal{U}_{st}.$$

(3.8)

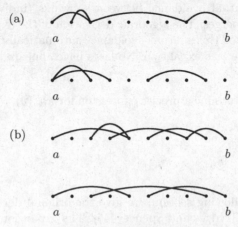

Fig. 3.1. Graphs in which an edge st is represented by an arc joining s and t. The graphs in (a) are not connected, whereas the graphs in (b) are connected.

Note that $\mathcal{B}[a,b]$ contains the graph with no edges, so our convention that $K[a,a] = 1$ is consistent with the standard convention that an empty product is equal to 1.

Exercise 3.3. Prove (3.8).

We set $J[a,a] = 1$, and for $a < b$ we define a quantity analogous to $K[a,b]$, but with the sum over graphs restricted to connected graphs:

$$J[a,b] = \sum_{\Gamma \in \mathcal{G}[a,b]} \prod_{st \in \Gamma} \mathcal{U}_{st}. \tag{3.9}$$

Lemma 3.4. *For any $a < b$,*

$$K[a,b] = K[a+1,b] + \sum_{j=a+1}^{b} J[a,j]K[j,b]. \tag{3.10}$$

Proof. The contribution to the sum on the right hand side of (3.8) due to all graphs Γ for which a is not in an edge is exactly $K[a+1,b]$. To resum the contribution due to the remaining graphs, we proceed as follows. If Γ does contain an edge containing a, let $j(\Gamma)$ be the largest value of j such that the set of edges in Γ with both ends in the interval $[a,j]$ forms a connected graph on $[a,j]$. Then the sum over Γ factorizes into sums over connected graphs on $[a,j]$ and arbitrary graphs on $[j,b]$, and resummation of the latter gives

$$K[a,b] = K[a+1,b] + \sum_{j=a+1}^{b} \sum_{\Gamma \in \mathcal{G}[a,j]} \prod_{st \in \Gamma} \mathcal{U}_{st} \, K[j,b], \tag{3.11}$$

which with (3.9) proves the lemma. ∎

Let

$$c_n(x) = \sum_{\omega \in \mathcal{W}_n(x)} K[0,n] = \sum_{\omega \in \mathcal{W}_n(x)} \prod_{0 \le s < t \le n} (1 + \mathcal{U}_{st}(\omega)), \tag{3.12}$$

a generalization of (2.2). It is simplest if we assume that $\mathcal{U}_{st}(\omega)$ is invariant under spatial translation of ω, and under an equal shift of each of s,t and the time parameter of ω, and we make this assumption. Note that (2.1) obeys the assumption. We substitute (3.10) into (3.12). A key point is that in the last term of (3.10) the portion of the walk from time j onwards is independent of the portion up to time j. Let

$$\pi_m(x) = \sum_{\omega \in \mathcal{W}_m(0,x)} J[0,m]. \tag{3.13}$$

Then for $n \ge 1$, we obtain

$$c_n(x) = (|\Omega|D * c_{n-1})(x) + \sum_{m=1}^{n} (\pi_m * c_{n-m})(x), \tag{3.14}$$

as in (3.5).[1] To obtain a more useful representation of π_m than (3.13), we perform a resummation of (3.13) using the notion of laces.

3.3 Laces and Resummation

Definition 3.5. A lace *is a minimally connected graph, i.e., a connected graph for which the removal of any edge would result in a disconnected graph. The set of laces on* $[a,b]$ *is denoted by* $\mathcal{L}[a,b]$, *and the set of laces on* $[a,b]$ *which consist of exactly* N *edges is denoted* $\mathcal{L}^{(N)}[a,b]$.

We write $L \in \mathcal{L}^{(N)}[a,b]$ as $L = \{s_1 t_1, \ldots, s_N t_N\}$, with $s_l < t_l$ for each l. The fact that L is a lace is equivalent to a certain ordering of the s_l and t_l. For $N = 1$, we simply have $a = s_1 < t_1 = b$. For $N \geq 2$, $L \in \mathcal{L}^{(N)}[a,b]$ if and only if

$$a = s_1 < s_2, \quad s_{l+1} < t_l \leq s_{l+2} \quad (l = 1, \ldots, N-2), \quad s_N < t_{N-1} < t_N = b$$
(3.15)

(for $N = 2$ the vacuous middle inequalities play no role); see Fig. 3.2. Thus L divides $[a,b]$ into $2N - 1$ subintervals:

$$[s_1, s_2], [s_2, t_1], [t_1, s_3], [s_3, t_2], \ldots, [s_N, t_{N-1}], [t_{N-1}, t_N].$$
(3.16)

Of these, intervals number 3, 5, ..., $(2N-3)$ can have zero length for $N \geq 3$, whereas all others have length at least 1.

Exercise 3.6. Prove that (3.15) characterizes laces.

Given a connected graph $\Gamma \in \mathcal{G}[a,b]$, the following prescription associates to Γ a unique lace $\mathsf{L}_\Gamma \subset \Gamma$: The lace L_Γ consists of edges $s_1 t_1, s_2 t_2, \ldots$, with $t_1, s_1, t_2, s_2, \ldots$ determined, in that order, by

$$t_1 = \max\{t : at \in \Gamma\}, \quad s_1 = a,$$

$$t_{i+1} = \max\{t : \exists s < t_i \text{ such that } st \in \Gamma\}, \quad s_{i+1} = \min\{s : st_{i+1} \in \Gamma\}.$$

The procedure terminates when $t_{i+1} = b$. Given a lace L, the set of all edges $st \notin L$ such that $\mathsf{L}_{L \cup \{st\}} = L$ is denoted $\mathcal{C}(L)$. Edges in $\mathcal{C}(L)$ are said to be *compatible* with L. Fig. 3.3 illustrates these definitions.

Exercise 3.7. Show that $\mathsf{L}_\Gamma = L$ if and only if L is a lace, $L \subset \Gamma$, and $\Gamma \setminus L \subset \mathcal{C}(L)$.

---

[1] For $m = 1$, there is a single connected graph $\{01\}$, and when U_{st} is given by (2.1) we have $\pi_1(x) = \sum_{\omega \in \mathcal{W}_1(0,x)} U_{01}(\omega) = 0$, since it is always the case that $\omega(0) \neq \omega(1)$. Thus the sum over m in (3.14) can be started at $m = 2$ in this case.

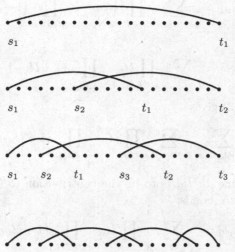

Fig. 3.2. Laces in $\mathcal{L}^{(N)}[a,b]$ for $N = 1, 2, 3, 4$, with $s_1 = a$ and $t_N = b$.

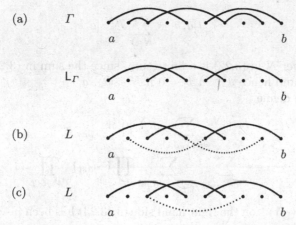

Fig. 3.3. (a) A connected graph Γ and its associated lace $L = \mathsf{L}_\Gamma$. (b) The dotted edges are compatible with the lace L. (c) The dotted edge is not compatible with the lace L.

The sum over connected graphs in (3.9) can be performed by first summing over all laces and then, given a lace, summing over all connected graphs associated to that lace by the above prescription. This gives

$$J[a,b] = \sum_{L \in \mathcal{L}[a,b]} \prod_{st \in L} \mathcal{U}_{st} \sum_{\Gamma : \mathsf{L}_\Gamma = L} \prod_{s't' \in \Gamma \setminus L} \mathcal{U}_{s't'}. \qquad (3.17)$$

But, writing $\Gamma' = \Gamma \setminus L$, it follows from Exercise 3.7 that

$$\sum_{\Gamma: L_\Gamma = L} \prod_{s't' \in \Gamma \setminus L} \mathcal{U}_{s't'} = \sum_{\Gamma' \subset \mathcal{C}(L)} \prod_{s't' \in \Gamma'} \mathcal{U}_{s't'} = \prod_{s't' \in \mathcal{C}(L)} (1 + \mathcal{U}_{s't'}). \qquad (3.18)$$

Therefore,

$$J[a, b] = \sum_{L \in \mathcal{L}[a,b]} \prod_{st \in L} \mathcal{U}_{st} \prod_{s't' \in \mathcal{C}(L)} (1 + \mathcal{U}_{s't'}). \qquad (3.19)$$

Inserting this in (3.13) gives

$$\pi_m(x) = \sum_{\omega \in \mathcal{W}_m(0,x)} \sum_{L \in \mathcal{L}[0,m]} \prod_{st \in L} \mathcal{U}_{st} \prod_{s't' \in \mathcal{C}(L)} (1 + \mathcal{U}_{s't'}). \qquad (3.20)$$

For $a < b$ we define $J^{(N)}[a, b]$ to be the contribution to (3.17) from laces consisting of exactly N bonds:

$$J^{(N)}[a, b] = \sum_{L \in \mathcal{L}^{(N)}[a,b]} \prod_{st \in L} \mathcal{U}_{st} \prod_{s't' \in \mathcal{C}(L)} (1 + \mathcal{U}_{s't'}). \qquad (3.21)$$

For the special case in which $\mathcal{U}_{st}$ is given by (2.1), each term in the above sum is either 0 or $(-1)^N$. By (3.17) and (3.21),

$$J[a, b] = \sum_{N=1}^{\infty} J^{(N)}[a, b]. \qquad (3.22)$$

The sum over N in (3.22) is a finite sum, since the sum in (3.21) is empty for $N > b - a$ and hence $J^{(N)}[a, b] = 0$ if $N > b - a$.

Now we define

$$\pi_m^{(N)}(x) = (-1)^N \sum_{\omega \in \mathcal{W}_m(x)} J^{(N)}[0, m]$$

$$= \sum_{\omega \in \mathcal{W}_m(x)} \sum_{L \in \mathcal{L}^{(N)}[0,m]} \prod_{st \in L} (-\mathcal{U}_{st}) \prod_{s't' \in \mathcal{C}(L)} (1 + \mathcal{U}_{s't'}). \qquad (3.23)$$

The factor $(-1)^N$ on the right hand side of (3.23) has been inserted to arrange that

$$\pi_m^{(N)}(x) \geq 0 \quad \text{for all } N, m, x \qquad (3.24)$$

when $\mathcal{U}_{st}$ is given by U_{st} of (2.1). By (3.13), (3.22) and (3.23),

$$\pi_m(x) = \sum_{N=1}^{\infty} (-1)^N \pi_m^{(N)}(x). \qquad (3.25)$$

For the special case in which $\mathcal{U}_{st}$ is given by (2.1), walks making a nonzero contribution to (3.23) are constrained to have the topology indicated in Fig. 3.4. In the figure, for $\prod_{s't' \in \mathcal{C}(L)} (1 + \mathcal{U}_{s't'}) \neq 0$, each of the $2N - 1$ subwalks

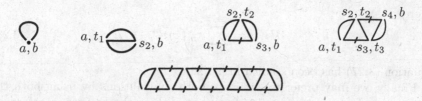

Fig. 3.4. Self-intersections required for a walk ω with $\prod_{st\in L} U_{st}(\omega) \neq 0$, with U_{st} given by (2.1), for the laces with $N = 1, 2, 3, 4$ bonds depicted in Fig. 3.2. The picture for $N = 11$ is also shown.

must be a self-avoiding walk, and in addition there must be mutual avoidance between some (but not all) of the subwalks. The number of loops (faces excluding the "outside" face) in a diagram is equal to the number of edges in the corresponding lace. The lines which are slashed correspond to subwalks which may consist of zero steps, but the others correspond to subwalks consisting of at least one step. This gives an interpretation of $\pi_m^{(N)}$ identical to that obtained in Sect. 3.1, but here there is the advantage that explicit formulas keep track of the mutual avoidance between subwalks.

It is sometimes convenient to modify the definitions of "connected graph" and "lace," and we will do so in Sect. 8.1. A more general theory of laces is developed and applied in [124, 126], for the analysis of networks of mutually-avoiding self-avoiding walks. See also [125] for an application of the more general theory to lattice trees.

3.4 Transformations

Equation (3.14) involves convolution in both space and time. It has been studied in this form in [29], via fixed point methods.

It is tempting to use transformations to eliminate one or both of these convolutions. We can eliminate the convolution in space if we take the Fourier transform (1.6). For $n \geq 1$, this gives

$$\hat{c}_n(k) = |\Omega|\hat{D}(k)\hat{c}_{n-1}(k) + \sum_{m=1}^{n} \hat{\pi}_m(k)\hat{c}_{n-m}(k). \tag{3.26}$$

Conditions are given in [120] which ensure that solutions of (3.26) have Gaussian asymptotics, via an analysis based on induction on n.

We may instead prefer to eliminate the convolution in time, by going to generating functions. Using (2.18) and (3.14), this gives

$$G_z(x) = \delta_{0,x} + \sum_{n=1}^{\infty} c_n(x)z^n$$
$$= \delta_{0,x} + z|\Omega|(D * G_z)(x) + (\Pi_z * G_z)(x), \tag{3.27}$$

where

$$\Pi_z(x) = \sum_{m=1}^{\infty} \pi_m(x) z^m. \tag{3.28}$$

Equation (3.27) has been studied in [90, 91].

Finally, we may prefer to eliminate both convolutions by using both the Fourier transform and generating functions. Taking the Fourier transform of (3.27) gives

$$\hat{G}_z(k) = 1 + z|\Omega|\hat{D}(k)\hat{G}_z(k) + \hat{\Pi}_z(k)\hat{G}_z(k), \tag{3.29}$$

which can be solved to give

$$\hat{G}_z(k) = \frac{1}{1 - z|\Omega|\hat{D}(k) - \hat{\Pi}_z(k)}. \tag{3.30}$$

Equation (3.30) has been the point of departure for several studies of the self-avoiding walk, and we will work with (3.30) in Chap. 5.

Exercise 3.8. The memory-2 walk is the walk with $\mathcal{U}_{st} = U_{st}$ if $t - s \leq 2$, and otherwise $\mathcal{U}_{st} = 0$. This is a random walk with no immediate reversals. Suppose that $0 \notin \Omega \subset \mathbb{Z}^d$ is finite and invariant under the symmetries of the lattice.
(a) What is the value of $\hat{c}_n(0)$, the number of n-step memory-2 walks? (Calculation is not required.)
(b) Prove that for the memory-2 walk, for $m \geq 2$,

$$\pi_m(x) = \begin{cases} -|\Omega|\delta_{x,0} & \text{if } m \text{ is even} \\ I[x \in \Omega] & \text{if } m \text{ is odd}. \end{cases}$$

(c) Suppose that $|\Omega| > 2$. Show that the mean-square displacement for the memory-two walk is given by

$$\sigma^2 \left[\left(\frac{1+\delta}{1-\delta} \right) n - \frac{2\delta(1-\delta^n)}{(1-\delta)^2} \right] \sim \left(\frac{\sigma^2|\Omega|}{|\Omega|-2} \right) n,$$

where $\sigma^2 = \sum_x |x|^2 D(x)$ is the variance of D and $\delta = (|\Omega| - 1)^{-1}$. One approach[2] is to use (3.26) to compute $\nabla^2 \hat{c}_n(0)$. This problem goes back a long way [18, 63, 72].
(d) Show that for the memory-two walk,

$$\hat{G}_z(k) = \frac{1 - z^2}{1 + (|\Omega| - 1)z^2 - z|\Omega|\hat{D}(k)}$$

---

[2] Verification of the formula by induction seems an unsatisfactory solution, since it requires prior knowledge of the formula.

(compare Exercise 2.2 for $d = 1$). This formula was used to compute the mean-square displacement via contour integration in [158, Sect. 5.3].

The memory-τ walk is the walk with $\mathcal{U}_{st} = U_{st}$ if $t - s \leq \tau$, and otherwise $\mathcal{U}_{st} = 0$. Finite-memory walks played an important role in the original analysis of the lace expansion in [45], but will not concern us further here. For a study of generating functions of the number of memory-τ walks, for $\tau \leq 8$, see [171].

Diagrammatic Estimates for the Self-Avoiding Walk

The difficulty in analyzing the lace expansion is to understand the function $\pi_m(x)$, or one of its transforms. In this chapter, we will prove estimates for the Fourier transform $\hat{\Pi}_z(k)$ of the generating function $\Pi_z(x) = \sum_{m=1}^{\infty} \pi_m(x) z^m$. Related estimates of one sort or another have been used in every analysis of the lace expansion for the self-avoiding walk. Throughout this chapter, we use the notation of Sect. 3.2, and we take $\mathcal{U}_{st}$ to be given by (2.1), i.e.,

$$\mathcal{U}_{st} = U_{st} = \begin{cases} -1 \text{ if } \omega(s) = \omega(t) \\ 0 \text{ if } \omega(s) \neq \omega(t). \end{cases} \tag{4.1}$$

We also assume that our walks take steps in a finite set Ω which is invariant under the symmetries of $\mathbb{Z}^d$, namely permutation of coordinates and replacement of any coordinate x_i by $-x_i$.

We will obtain estimates for $\sum_{x \in \mathbb{Z}^d} \Pi_z^{(N)}(x)$, which is an upper bound for $|\hat{\Pi}_z^{(N)}(k)|$, and for $\sum_{x \in \mathbb{Z}^d} [1 - \cos(k \cdot x)] \Pi_z^{(N)}(x)$, which is an upper bound for $\hat{\Pi}_z^{(N)}(0) - \hat{\Pi}_z^{(N)}(k)$. To motivate the latter, let $\hat{F}_z(k) = 1/\hat{G}_z(k)$, and note from (3.30) that

$$\hat{G}_z(k) = \frac{1}{\hat{F}_z(0) + [\hat{F}_z(k) - \hat{F}_z(0)]}$$
$$= \frac{1}{\hat{F}_z(0) + z|\Omega|[1 - \hat{D}(k)] + [\hat{\Pi}_z(0) - \hat{\Pi}_z(k)]}. \tag{4.2}$$

Our estimate for $\sum_{x \in \mathbb{Z}^d} [1 - \cos(k \cdot x)] \Pi_z^{(N)}(x)$ will ultimately allow us to compare the terms $[\hat{\Pi}_z(0) - \hat{\Pi}_z(k)]$ and $z|\Omega|[1 - \hat{D}(k)]$ in the denominator.

4.1 The Diagrammatic Estimates

Recall from (3.23) and (3.25) that $\pi_m(x) = \sum_{N=1}^{\infty} (-1)^N \pi_m^{(N)}(x)$, with

$$\pi_m^{(N)}(x) = \sum_{\omega \in \mathcal{W}_m(x)} \sum_{L \in \mathcal{L}^{(N)}[0,m]} \prod_{st \in L} (-U_{st}) \prod_{s't' \in \mathcal{C}(L)} (1 + U_{s't'}). \tag{4.3}$$

For $z \geq 0$, we define the non-negative generating function

$$\Pi_z^{(N)}(x) = \sum_{m=2}^{\infty} \pi_m^{(N)}(x) z^m. \tag{4.4}$$

In the above series, we omit the term involving $\pi_1^{(N)}(x)$ because it is always zero, since the only lace on $[0,1]$ is $L = \{01\}$, and $U_{01} = 0$ since a walk cannot be at the same place at consecutive times. By (3.28), we have

$$\Pi_z(x) = \sum_{N=1}^{\infty} (-1)^N \Pi_z^{(N)}(x). \tag{4.5}$$

Let

$$H_z(x) = G_z(x) - \delta_{0,x} = \sum_{n=1}^{\infty} c_n(x) z^n. \tag{4.6}$$

The following theorem gives bounds on $\Pi_z^{(N)}$ in terms of norms of G_z and H_z.

Theorem 4.1. *For all $z \geq 0$,*

$$\sum_{x \in \mathbb{Z}^d} \Pi_z^{(1)}(x) \leq z|\Omega| \, \|H_z\|_{\infty} \tag{4.7}$$

and

$$\sum_{x \in \mathbb{Z}^d} [1 - \cos(k \cdot x)] \Pi_z^{(1)}(x) = 0. \tag{4.8}$$

For $z \geq 0$ and $N \geq 2$,

$$\sum_{x \in \mathbb{Z}^d} \Pi_z^{(N)}(x) \leq \|H_z\|_{\infty} \|H_z * G_z\|_{\infty}^{N-1}, \tag{4.9}$$

and

$$\sum_{x \in \mathbb{Z}^d} [1 - \cos(k \cdot x)] \Pi_z^{(N)}(x) \tag{4.10}$$

$$\leq (N+1)\lfloor N/2 \rfloor \|[1 - \cos(k \cdot x)] H_z(x)\|_{\infty} \|H_z * G_z\|_{\infty}^{N-1}.$$

We refer to the bounds of Theorem 4.1 as diagrammatic estimates, as they are inspired by the diagrams of Fig. 3.4. Moreover, the diagrams themselves provide a pictorial representation of the bounds, and have the dual interpretation of depicting both walk trajectories that contribute to $\Pi_z^{(N)}(x)$ as well as upper bounds on these quantities. See Fig. 4.1.[1]

---

[1] Fig. 4.1 shows a slight improvement of (4.9) (as $H_z \leq G_z$) and it is possible to prove the improvement, but we will only prove (4.9), which suffices for our needs.

Fig. 4.1. Depiction of the estimate $\sum_x \Pi_z^{(4)}(x) \leq \|H_z * H_z\|_\infty \|H_z * G_z\|_\infty^2 \|H_z\|_\infty$ by decomposition of the diagram for $\Pi_z^{(4)}$.

In applying Theorem 4.1, we will use the estimate

$$\|H_z * G_z\|_\infty = \|H_z + (H_z * H_z)\|_\infty \leq \|H_z\|_\infty + \|H_z\|_2^2, \qquad (4.11)$$

using the triangle and Cauchy–Schwarz inequalities in the last step. Since $G_z(0) = 1$, it follows from (4.6) and (2.30) that

$$\|H_z\|_2^2 = \|G_z\|_2^2 - 1 = B(z) - 1. \qquad (4.12)$$

To control the sum over N in (4.5) using (4.9), our method in Chap. 5 will require, in particular, that $B(z_c) - 1$ be small. This is a restrictive form of the bubble condition of Sect. 2.2.

4.2 Proof of the Diagrammatic Estimates

In this section, we prove Theorem 4.1.

4.2.1 Proof of (4.7)–(4.8)

The estimates (4.7)–(4.8) are easy, and we prove them first. Since the unique lace on $[0, m]$ consisting of a single bond is simply the bond $0m$, it follows from (3.23) that

$$\pi_m^{(1)}(x) = \delta_{0,x} \sum_{\omega \in \mathcal{W}_m(0,0)} \prod_{s't' \in \mathcal{C}(0m)} (1 + U_{s't'}). \qquad (4.13)$$

There is no 1-step walk from 0 to 0, so this is nonzero only for $m \geq 2$. Since $\mathcal{C}(0m) \supset \mathcal{B}[0, m-1]$ for $m \geq 2$, it follows from (4.1) that

$$0 \leq \pi_m^{(1)}(x) \leq \delta_{0,x} \sum_{\omega \in \mathcal{W}_m(0,0)} K[0, m-1]$$
$$= \delta_{0,x} \sum_{y \in \Omega} c_{m-1}(y). \qquad (4.14)$$

Therefore, after multiplying by z^m and summing over $m \geq 2$, we obtain

$$0 \leq \Pi_z^{(1)}(x) \leq \delta_{0,x} z \sum_{y \in \Omega} H_z(y), \qquad (4.15)$$

which immediately implies (4.7)–(4.8).

4.2.2 The Diagrams

In preparation for the proof of (4.9)–(4.10), we now prove a preliminary estimate on $\Pi_z^{(N)}(x)$.

For $N \geq 2$, we define $p_m^{(N)}(x, y)$ inductively as follows. Let

$$p_m^{(2)}(x, y) = \sum_{0 < i < j \leq m} c_i(x)c_{j-i}(x)c_{m-j}(y) \quad (m \geq 2), \qquad (4.16)$$

$$a_m(u, v, x, y) = \delta_{v,x} \sum_{l=1}^{m} c_l(u - v)c_{m-l}(y - u) \quad (m \geq 1). \qquad (4.17)$$

For $m = 0, 1$, we set $p_m^{(2)}(x, y) = 0$, and we set $a_0(u, v, x, y) = 0$. For $N \geq 3$, let

$$p_m^{(N)}(x, y) = \sum_{u,v \in \mathbb{Z}^d} \sum_{i=0}^{m} p_i^{(N-1)}(u, v)a_{m-i}(u, v, x, y) \quad (m \geq 2). \qquad (4.18)$$

For $N \geq 2$, we also define the generating functions

$$P_z^{(N)}(x, y) = \sum_{m=2}^{\infty} p_m^{(N)}(x, y)z^m, \qquad (4.19)$$

$$A_z(u, v, x, y) = \sum_{m=1}^{\infty} a_m(u, v, x, y)z^m$$
$$= \delta_{v,x}H_z(u - v)G_z(y - u). \qquad (4.20)$$

It follows from (4.18) that, for $N \geq 3$,

$$P_z^{(N)}(x, y) = \sum_{u,v \in \mathbb{Z}^d} P_z^{(N-1)}(u, v)A_z(u, v, x, y). \qquad (4.21)$$

The diagrammatic representations for $P_z^{(N)}(x, y)$ shown in Fig. 4.2 are closely related to the diagrams appearing in Fig. 3.4.

Proposition 4.2. *For $N \geq 2$, $m \geq 2$, and $z \geq 0$,*

$$\pi_m^{(N)}(x) \leq p_m^{(N)}(x, x) \qquad (4.22)$$

and

$$\Pi_z^{(N)}(x) \leq P_z^{(N)}(x, x). \qquad (4.23)$$

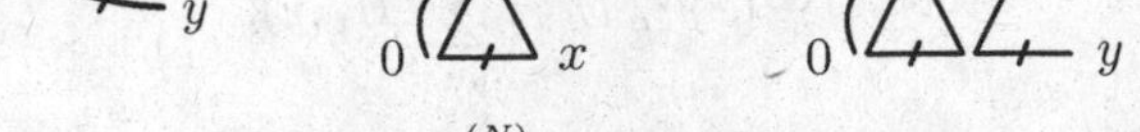

Fig. 4.2. $P_z^{(N)}(x, y)$ for $N = 2, 3, 4$.

Proof. We prove the first inequality, as the second then follows immediately from (4.4).

For $N \geq 2$ we write a lace $L \in \mathcal{L}^{(N-1)}[0,j]$ as $\{s_1 t_1, \ldots, s_{N-1} t_{N-1}\}$, with $s_1 = 0$ and $t_{N-1} = j$. For $N \geq 2$, we define

$$J_x^{(N)}[0,m] \tag{4.24}$$

$$= \sum_{j=0}^{m} K[j,m] \sum_{L \in \mathcal{L}^{(N-1)}[0,j]} \sum_{i=t_{N-2}}^{j-1} \delta_{x,\omega(i)} \prod_{st \in L} (-U_{st}) \prod_{s't' \in \mathcal{C}(L)} (1 + U_{s't'}),$$

where we set $t_0 = 1$ when $N = 2$. We first show that, for every ω and every $N \geq 2$,

$$0 \leq (-1)^N J^{(N)}[0,m] \leq J_{\omega(m)}^{(N)}[0,m]. \tag{4.25}$$

The first inequality is immediate, and we concentrate on the second. For this, comparing with (3.21), L in (4.24) corresponds to $L\backslash\{s_N t_N\}$ in (3.21), and i and j of (4.24) correspond to s_N and t_{N-1} of the lace L in (3.21). The set of compatible edges in (3.21) contains $\mathcal{C}(s_1 t_1, \ldots, s_{N-1} t_{N-1}) \cup \mathcal{B}[t_{N-1}, m]$, and omitting factors in the product over $s't'$ in (3.21) can only increase the product. When $x = \omega(m)$, the factor $\delta_{x,\omega(i)}$ is $-U_{i,m} = -U_{s_N t_N}$. This leads to (4.25).

For $N \geq 2$, we define

$$\pi_m^{(N)}(x,y) = \sum_{\omega \in \mathcal{W}_m(0,y)} J_x^{(N)}[0,m]. \tag{4.26}$$

It follows from (3.23) and (4.25) that

$$\pi_m^{(N)}(x) \leq \pi_m^{(N)}(x,x). \tag{4.27}$$

We will show that

$$\pi_m^{(N)}(x,y) \leq p_m^{(N)}(x,y) \qquad (N \geq 2), \tag{4.28}$$

which then gives the proposition.

The proof of (4.28) is by induction on N. We begin the induction with the case $N = 2$. In this case, the sum over L in (4.24) consists of the single term $L = \{0j\}$. Since $\mathcal{C}(0j) \supset \mathcal{B}[0,i] \cup \mathcal{B}[i,j]$ for $0 < i < j$, using symmetry we obtain

$$\pi_m^{(2)}(x,y) \leq \sum_{0 < i < j < m} \sum_{\omega \in \mathcal{W}_m(0,y)} K[0,i]\delta_{x,\omega(i)} K[i,j]\delta_{0,\omega(j)} K[j,m]$$

$$= \sum_{0 < i < j < m} c_i(x) c_{j-i}(x) c_{m-j}(y)$$

$$\leq p_m^{(2)}(x,y). \tag{4.29}$$

To advance the induction, we fix $N \geq 3$ and assume that (4.28) holds for $N - 1$. We replace the factor $-U_{s_{N-1}t_{N-1}}$ in the first product of (4.24) by

$$-U_{s_{N-1}t_{N-1}} = \delta_{\omega(s_{N-1}),\omega(t_{N-1})} = \sum_{u \in \mathbb{Z}^d} \delta_{\omega(s_{N-1}),u}\delta_{u,\omega(t_{N-1})}. \qquad (4.30)$$

Given $L \in \mathcal{L}^{(N-1)}[0,j]$, let $L' = L\backslash\{s_{N-1}t_{N-1}\}$. For $t_{N-2} \leq i < j = t_{N-1}$, we then have $\mathcal{C}(L) \supset \mathcal{C}(L') \cup \mathcal{B}[t_{N-2},i] \cup \mathcal{B}[i,j]$. Using (4.30), we conclude from (4.24) that

$$J_x^{(N)}[0,m] \leq \sum_u \sum_{0 \leq i < j \leq m} J_u^{(N-1)}[0,i]K[i,j]\delta_{\omega(i),x}K[j,m]\delta_{u,\omega(j)}. \qquad (4.31)$$

Therefore, recalling (4.17) and using the induction hypothesis,

$$\pi_m^{(N)}(x,y) \leq \sum_u \sum_{0 \leq i < j \leq m} \pi_i^{(N-1)}(u,x)c_{j-i}(u-x)c_{m-j}(y-u)$$

$$\leq \sum_{u,v} \sum_{i=0}^m p_i^{(N-1)}(u,v)a_{m-i}(u,v,x,y)$$

$$= p_m^{(N)}(x,y). \qquad (4.32)$$

This completes the proof.   ∎

Exercise 4.3. Convince yourself that (4.31) holds.

4.2.3 Proof of (4.9)–(4.10)

We prove two lemmas, and combine them with Proposition 4.2 to obtain (4.9)–(4.10).

We define the operators

$$(\mathcal{M}_z f)(x) = H_z(x)f(x), \qquad (4.33)$$

$$(\mathcal{H}_z f)(x) = (H_z * f)(x), \qquad (4.34)$$

$$(\mathcal{H}_z' f)(x) = (G_z * f)(x). \qquad (4.35)$$

Lemma 4.4. *For $N \geq 2$ and $z \geq 0$,*

$$\sum_x P_z^{(N)}(x, x+y) = \left[(\mathcal{H}_z'\mathcal{M}_z)^{N-1}H_z\right](y). \qquad (4.36)$$

Proof. The proof is by induction on N. For $N = 2$, we conclude from (4.16) that

$$\sum_x P_z^{(2)}(x, x+y) = \sum_x H_z(x)^2 G_z(x+y)$$

$$= \sum_x H_z(x)^2 G_z(y-x) = [(\mathcal{H}_z'\mathcal{M}_z)H_z](y), \qquad (4.37)$$

using $H_z(-x) = H_z(x)$ in the second step.

To advance the induction, we assume that (4.36) holds for $N-1$. By (4.20)–(4.21),

$$\sum_x P_z^{(N)}(x,x+y) = \sum_{x,u,v} P_z^{(N-1)}(u,v)A_z(u,v,x,x+y)$$

$$= \sum_{u,v} P_z^{(N-1)}(u,v)H_z(u-v)G_z(v-u+y)$$

$$= \sum_w \left(\sum_u P_z^{(N-1)}(u,u+w)\right) H_z(-w)G_z(w+y)$$

$$= \sum_w \left(\sum_u P_z^{(N-1)}(u,u+w)\right) H_z(w)G_z(y-w), \qquad (4.38)$$

where in the last step we replaced w by $-w$ and used the fact that the first factor is unchanged by this replacement (see Exercise 4.5 below). Writing $F(w)$ for the first factor, the above is equal to

$$(G_z * H_z F)(y) = (\mathcal{H}_z'\mathcal{M}_z F)(y). \qquad (4.39)$$

We then apply the induction hypothesis to complete the proof. ∎

Exercise 4.5. Prove the identity $\sum_u P_z^{(N)}(u,u+w) = \sum_u P_z^{(N)}(u,u-w)$ used in (4.38).

The combination of Proposition 4.2 with Lemma 4.4 gives

$$\sum_x \Pi_z^{(N)}(x) \le \sum_x P_z^{(N)}(x,x) = \left[(\mathcal{H}_z'\mathcal{M}_z)^{N-1}H_z\right](0). \qquad (4.40)$$

Note that for $N=2$ the upper bound can be replaced by $[(\mathcal{H}_z\mathcal{M}_z)^{N-1}H_z](0)$, which is equal to the above right hand side in this special case. The right hand side of (4.40) can be estimated using the following lemma.

Lemma 4.6. *Given non-negative even functions $f_0, f_1, \ldots, f_{2M}$ on $\mathbb{Z}^d$, define $\mathcal{H}_j$ and $\mathcal{M}_j$ to be respectively the operations of convolution with f_{2j} and multiplication by f_{2j-1}, for $j=1,\ldots,M$. Then for any $k \in \{0,\ldots,2M\}$,*

$$\|\mathcal{H}_M\mathcal{M}_M\cdots\mathcal{H}_1\mathcal{M}_1 f_0\|_\infty \le \|f_k\|_\infty \prod \|f_j * f_{j'}\|_\infty, \qquad (4.41)$$

where the product is over disjoint consecutive pairs j, j' taken from the set $\{0,\ldots,2M\} \setminus \{k\}$ (e.g., for $k=3$ and $M=3$, the product has factors with j, j' equal to $0,1; 2,4; 5,6$).

Proof. The proof is by induction on M. The desired result for $M = 1$ is a consequence of the elementary estimates

$$\sum_y f_2(x-y)f_1(y)f_0(y) \leq \begin{cases} \|f_0\|_\infty \|f_1 * f_2\|_\infty \\ \|f_1\|_\infty \|f_0 * f_2\|_\infty \\ \|f_2\|_\infty \|f_0 * f_1\|_\infty, \end{cases} \tag{4.42}$$

where for the last of these inequalities we used the fact that $\sum_y f_1(y)f_0(y) = (f_0 * f_1)(0)$ for even f_0. To advance the induction, we assume that (4.41) holds for $1, \ldots, M-1$. We write the function inside the norm on the left hand side of (4.41) as $\mathcal{H}_M \mathcal{M}_M F_{M-1}$, with $F_l = \mathcal{H}_l \mathcal{M}_l \cdots \mathcal{H}_1 \mathcal{M}_1 f_0$, and estimate its infinity norm using the result for $M = 1$. If we associate the infinity norm to F_{M-1}, an estimate of the form (4.41) follows from the induction hypothesis, for any $k \leq M - 1$. It remains to show that the infinity norm can also be associated to f_{2M} or f_{2M-1}.

We show this for the latter, and the former is similar. Applying the $M = 1$ case to $\mathcal{H}_M \mathcal{M}_M F_{M-1}$ gives an upper bound $\|f_{2M-1}\|_\infty \|f_{2M} * F_{M-1}\|_\infty$. Let $\widetilde{\mathcal{H}}_{M-1}$ denote convolution by $f_{2M} * f_{2M-2}$, so that

$$f_{2M} * F_{M-1} = \widetilde{\mathcal{H}}_{M-1} \mathcal{M}_{M-1} F_{M-2} \tag{4.43}$$

(with $F_0 = f_0$). We apply the induction hypothesis to estimate the infinity norm of the right hand side, associating the infinity norm to $\widetilde{\mathcal{H}}_{M-1}$. This gives the desired estimate. ∎

Exercise 4.7. Give the details omitted at the end of the above proof, for the case in which the infinity norm is associated to f_{2M}.

Proof of (4.9)–(4.10). The bound (4.9) follows from (4.40) and Lemma 4.6.

It remains to prove (4.10). Fix $N \geq 2$. Our goal is to estimate

$$\sum_x [1 - \cos(k \cdot x)]\Pi_z^{(N)}(x) = \sum_{m=2}^\infty z^m \sum_x [1 - \cos(k \cdot x)]\pi_m^{(N)}(x). \tag{4.44}$$

To do so, we investigate how the argument leading to (4.40) is modified by the factor $[1 - \cos(k \cdot x)]$.

Because of the factor $\prod_{ij \in L}(-U_{ij})$ occurring in the definition of $\pi_m^{(N)}(x)$ (see (4.3)), a nonzero contribution occurs only for those ω for which $\omega(i) = \omega(j)$ for each edge $ij \in L$. Let I_j denote the jth time interval listed in (3.16) ($j = 1, \ldots, 2N - 1$), and let y_j denote the displacement performed on I_j by a walk ω contributing to $\pi_m^{(N)}(x)$. These displacements y_j correspond to the subwalk displacements in Figs. 3.4 and 4.2. The constraints that $\omega(i) = \omega(j)$ for all $ij \in L$, together with the subinterval structure (3.16), impose the constraints

$$y_1 + y_2 = 0, \quad \sum_{j=2p}^{2p+2} y_j = 0 \quad (p = 1, \ldots, N-2), \quad y_{2N-2} + y_{2N-1} = 0. \tag{4.45}$$

It can also be seen from this (see also Fig. 3.4) that the total displacement x is given by

$$x = \sum_{i=1}^{\lfloor N/2 \rfloor} y_{4i-1} = \sum_{i=1}^{\lceil N/2 \rceil} y_{4i-3} = -\sum_{i=1}^{N-1} y_{2i} \qquad (4.46)$$

(we need only the first equality).

Let $t = \sum_{j=1}^{J} t_j$. Taking the real part of the telescoping sum

$$1 - e^{it} = \sum_{j=1}^{J} [1 - e^{it_j}] e^{i\sum_{m=1}^{j-1} t_m} \qquad (4.47)$$

leads to the bound

$$1 - \cos t \le \sum_{j=1}^{J} [1 - \cos t_j] + \sum_{j=1}^{J} \sin t_j \sin \left(\sum_{m=1}^{j-1} t_m \right). \qquad (4.48)$$

It is a consequence of the identity $\sin(x+y) = \sin x \cos y + \cos x \sin y$ that $|\sin(x+y)| \le |\sin x| + |\sin y|$. Applying this recursively gives

$$1 - \cos t \le \sum_{j=1}^{J} [1 - \cos t_j] + \sum_{j=1}^{J} \sum_{m=1}^{j-1} |\sin t_j|| \sin t_m|. \qquad (4.49)$$

In the last term we use $|ab| \le (a^2 + b^2)/2$, and then $1 - \cos^2 a \le 2[1 - \cos a]$, to obtain

$$
\begin{aligned}
1 - \cos t &\le \sum_{j=1}^{J} [1 - \cos t_j] + \frac{1}{2} \sum_{j=1}^{J} \sum_{m=1}^{j-1} [\sin^2 t_j + \sin^2 t_m] \\
&\le \sum_{j=1}^{J} [1 - \cos t_j] + J \sum_{j=1}^{J} \sin^2 t_j \\
&= \sum_{j=1}^{J} [1 - \cos t_j] + J \sum_{j=1}^{J} [1 - \cos^2 t_j] \\
&\le (2J+1) \sum_{j=1}^{J} [1 - \cos t_j]. \qquad (4.50)
\end{aligned}
$$

We use the decomposition of x given by the first equality of (4.46), and apply (4.50) with $t = k \cdot x = \sum_{j=1}^{\lfloor N/2 \rfloor} k \cdot y_{4j-1}$, to obtain

$$1 - \cos(k \cdot x) \le (N+1) \sum_{j=1}^{\lfloor N/2 \rfloor} [1 - \cos(k \cdot y_{4j-1})]. \tag{4.51}$$

The modification of the upper bound (4.40) due to the factor $[1 - \cos(k \cdot y_{4j-1})]$ is simply to replace one of the factors H_z or G_z occurring in the right hand side by $[1 - \cos(k \cdot y_{4j-1})]H_z(y_{4j-1})$. Then we apply Lemma 4.6, associating the infinity norm to this particular factor, to obtain the desired estimate (4.10). ∎

5

Convergence for the Self-Avoiding Walk

In this chapter, we prove convergence of the lace expansion for the nearest-neighbour model in sufficiently high dimensions, and for sufficiently spread-out models in dimensions $d > 4$. As part of the proof, we will show that the critical bubble diagram $B(z_c)$ is finite in these cases, and hence, by Theorem 2.3, that the critical exponent γ exists and equals 1. This is restated in the following theorem.

Theorem 5.1. *The bubble condition $B(z_c) < \infty$ for the self-avoiding walk holds for the nearest-neighbour model in dimensions $d \geq d_0$, and for the spread-out model with $L \geq L_0(d)$ in dimensions $d > 4$, for some constants d_0 and $L_0(d)$. Thus the critical exponent γ exists and equals 1, in the sense that $\chi(z) \simeq (1 - z/z_c)^{-1}$ as $z \to z_c^-$.*

Remark 5.2. The conclusion of Theorem 5.1 can easily be improved to an asymptotic formula $\chi(z) \sim A(1 - z/z_c)^{-1}$ as $z \to z_c^-$. See Exercise 5.19 below.

Recall that $H_z(x) = G_z(x) - \delta_{0,x} = \sum_{n=1}^{\infty} c_n(x)z^n$, so that $\|H_z\|_2^2 = \|G_z\|_2^2 - 1 = B(z) - 1$. We will prove not just that the critical bubble diagram $B(z_c)$ is finite, but that in fact $\|H_{z_c}\|_2^2 = B(z_c) - 1$ is small, under the hypotheses of Theorem 5.1. As a preliminary, we first analyze some related issues for simple random walk.

5.1 Random-Walk Estimates

By the Parseval relation, $\|H_{z_c}\|_2^2 = \|\hat{H}_{z_c}\|_2^2 = \|\hat{G}_{z_c} - 1\|_2^2$. By (1.18), the random walk analogue of the latter is the integral

$$\int_{[-\pi,\pi]^d} \left| \frac{1}{[1 - \hat{D}(k)]} - 1 \right|^2 \frac{d^d k}{(2\pi)^d} = \int_{[-\pi,\pi]^d} \frac{\hat{D}(k)^2}{[1 - \hat{D}(k)]^2} \frac{d^d k}{(2\pi)^d}. \quad (5.1)$$

The following proposition shows that this integral is small under the hypotheses of Theorem 5.1. (We have already encountered a closely related integral in Exercise 1.7.)

Proposition 5.3. *Let $d > 4$. Then*

$$\int_{[-\pi,\pi]^d} \frac{\hat{D}(k)^2}{[1 - \hat{D}(k)]^2} \frac{\mathrm{d}^d k}{(2\pi)^d} \leq \beta, \qquad (5.2)$$

with $\beta = K(d-4)^{-1}$ (K a universal constant) for the nearest-neighbour model, and with $\beta = KL^{-d}$ (K dependent on d) for the spread-out model.

Proof. This is a calculus problem. For the nearest-neighbour model, a proof can be found in [158, Lemma A.3]. For the spread-out model, there is a proof in [158, Lemma A.5] but with a β which is larger by a factor $(\log L)^{d/2}$. We show here how the improvement can be achieved for the spread-out model.

It is shown in [120] that there are positive constants η, c_1 (independent of L) such that for all $k \in [-\pi, \pi]^d$,

$$1 - \hat{D}(k) \geq c_1 L^2 |k|^2 \qquad (\|k\|_\infty \leq L^{-1}), \qquad (5.3)$$
$$1 - \hat{D}(k) > \eta \qquad (\|k\|_\infty \geq L^{-1}). \qquad (5.4)$$

The integral $(2\pi)^{-d} \int_{[-\pi,\pi]^d} \hat{D}(k)^2 \mathrm{d}^d k$ is equal to $(D * D)(0)$, which is the probability of return to the origin after two steps, namely $|\Omega|^{-1}$. For $j \geq 4$ even, it follows from (5.3)–(5.4) that

$$\int_{[-\pi,\pi]^d} \hat{D}(k)^j \frac{\mathrm{d}^d k}{(2\pi)^d} \leq \int_{\mathbb{R}^d} e^{-c_1 j L^2 |k|^2} \frac{\mathrm{d}^d k}{(2\pi)^d}$$
$$+ \int_{[-\pi,\pi]^d} \hat{D}(k)^2 (1 - \eta)^{j-2} \frac{\mathrm{d}^d k}{(2\pi)^d}$$
$$\leq \mathrm{const} L^{-d} j^{-d/2} + |\Omega|^{-1}(1 - \eta)^{j-2}$$
$$\leq \mathrm{const} L^{-d} j^{-d/2}. \qquad (5.5)$$

For $j \geq 3$ odd,

$$(2\pi)^{-d} \int_{[-\pi,\pi]^d} |\hat{D}(k)|^j \mathrm{d}^d k \leq (2\pi)^{-d} \int_{[-\pi,\pi]^d} \hat{D}(k)^{j-1} \mathrm{d}^d k$$
$$\leq \mathrm{const} L^{-d} j^{-d/2}, \qquad (5.6)$$

applying the estimate for j even in the last step. Now we expand $[1 - \hat{D}(k)]^{-2}$ in (5.2) as $\sum_{j=1}^\infty j \hat{D}(k)^{j-1}$, and use the above estimates to see that the left hand side of (5.2) is bounded above by a multiple of L^{-d}, assuming $d > 4$. ∎

Exercise 5.4. Prove (5.2) for the nearest-neighbour model, with $\beta = K(d - 4)^{-1}$.

The following lemma notes some useful implications of (5.2). The left hand sides of (5.8)–(5.9) are respectively the random walk analogues of $\|G_{z_c}\|_2^2$ and of $\|[1 - \cos(k \cdot x)]G_{z_c}(x)\|_\infty = \|[1 - \cos(k \cdot x)]H_{z_c}(x)\|_\infty$ (cf. (4.10)).

Lemma 5.5. *If (5.2) holds, then for $z \in [0, 1/|\Omega|]$,*

$$\sup_{x \in \mathbb{Z}^d} D(x) \leq \beta, \tag{5.7}$$

$$\|C_z\|_2^2 \leq 1 + 3\beta, \tag{5.8}$$

$$\|[1 - \cos(k \cdot x)]C_z(x)\|_\infty \leq 5(1 + 3\beta)[1 - \hat{D}(k)]. \tag{5.9}$$

We first prove (5.7)–(5.8).

Proof of (5.7)–(5.8). The left hand side of (5.7) is simply $|\Omega|^{-1}$. Since the left hand side of (5.2) is at least $(2\pi)^{-d} \int_{[-\pi,\pi]^d} \hat{D}(k)^2 \mathrm{d}^d k = |\Omega|^{-1}$, the bound (5.7) follows.

For (5.8) (and also (5.9)), it suffices to consider $z = 1/|\Omega|$. We use the Parseval relation to rewrite the left hand side as $\|\hat{C}_{1/|\Omega|}\|_2^2$. By (1.18), this equals

$$\int_{[-\pi,\pi]^d} \frac{1}{[1 - \hat{D}(k)]^2} \frac{\mathrm{d}^d k}{(2\pi)^d}$$

$$= \int_{[-\pi,\pi]^d} \left(1 + 2\frac{\hat{D}(k)}{[1 - \hat{D}(k)]} + \frac{\hat{D}(k)^2}{[1 - \hat{D}(k)]^2}\right) \frac{\mathrm{d}^d k}{(2\pi)^d}$$

$$\leq \int_{[-\pi,\pi]^d} \left(1 + 3\frac{\hat{D}(k)^2}{[1 - \hat{D}(k)]^2}\right) \frac{\mathrm{d}^d k}{(2\pi)^d}, \tag{5.10}$$

by Exercise 5.6. The right hand side is at most $1 + 3\beta$, assuming (5.2). ∎

Exercise 5.6. Prove the inequality (5.10) by comparing the integrals of $\hat{D}[1 - \hat{D}]^{-1}$ and $\hat{D}^2[1 - \hat{D}]^{-2}$.

Before proving (5.9), we develop some useful preliminaries. We first note that

$$\sum_x \cos(k \cdot x)C_z(x)e^{il \cdot x} = \frac{1}{2}[\hat{C}_z(l + k) + \hat{C}_z(l - k)]. \tag{5.11}$$

Therefore, applying the general fact that $\|f\|_\infty \leq \|\hat{f}\|_1$, we obtain

$$\|[1 - \cos(k \cdot x)]C_z(x)\|_\infty \leq \|\hat{C}_z(l) - \frac{1}{2}[\hat{C}_z(l + k) + \hat{C}_z(l - k)]\|_1, \tag{5.12}$$

where the L^1 norm involves integration with respect to l, with k fixed. The expression $\hat{C}_z(l) - \frac{1}{2}(\hat{C}_z(l+k) + \hat{C}_z(l-k))$ is closely related to a sort of second derivative of $\hat{C}_z(l)$, and in general we make the abbreviation

$$-\frac{1}{2}\Delta_k\hat{A}(l) = \hat{A}(l) - \frac{1}{2}(\hat{A}(l+k) + \hat{A}(l-k)). \tag{5.13}$$

In this notation, (5.12) reads

$$\|[1 - \cos(k \cdot x)C_z(x)\|_\infty \le \frac{1}{2}\|\Delta_k\hat{C}_z(l)\|_1, \tag{5.14}$$

where the integration on the right hand side is with respect to l.

Lemma 5.7. *Suppose that* $a(-x) = a(x)$ *for all* $x \in \mathbb{Z}^d$, *and let*

$$\hat{A}(k) = \frac{1}{1 - \hat{a}(k)}. \tag{5.15}$$

Then for all $k, l \in [-\pi, \pi]^d$,

$$\frac{1}{2}|\Delta_k\hat{A}(l)| \le \frac{1}{2}[\hat{A}(l-k) + \hat{A}(l+k)]\hat{A}(l)[\hat{a}^{\mathrm{av}}(0) - \hat{a}^{\mathrm{av}}(k)] \tag{5.16}$$
$$+ 4\hat{A}(l-k)\hat{A}(l)\hat{A}(l+k)[\hat{a}^{\mathrm{av}}(0) - \hat{a}^{\mathrm{av}}(k)][\hat{a}^{\mathrm{av}}(0) - \hat{a}^{\mathrm{av}}(l)],$$

where $a^{\mathrm{av}}(x) = |a(x)|$.

Proof of (5.9). We use Lemma 5.7 to estimate the right hand side of (5.14), with $\hat{a}(k) = \hat{D}(k)$ and $\hat{A}(k) = \hat{C}_{1/|\Omega|}(k)$. Writing the latter simply as $\hat{C}(k)$, this gives

$$\frac{1}{2}|\Delta_k\hat{C}(l)| \le [1 - \hat{D}(k)]\left(\frac{1}{2}[\hat{C}(l-k) + \hat{C}(l+k)]\hat{C}(l) + 4\hat{C}(l-k)\hat{C}(l+k)\right). \tag{5.17}$$

Therefore, by the Cauchy-Schwarz inequality,

$$\frac{1}{2}\|\Delta_k\hat{C}\|_1 \le [1 - \hat{D}(k)]5\|\hat{C}\|_2^2, \tag{5.18}$$

and (5.9) follows from (5.8).   ∎

Proof of Lemma 5.7. Since a is even, $\hat{a}(l) = \sum_x a(x)\cos(l \cdot x)$. For such an a, we define

$$\hat{a}^{\cos}(l, k) = \sum_x a(x)\cos(l \cdot x)\cos(k \cdot x) = \frac{1}{2}[\hat{a}(l-k) + \hat{a}(l+k)], \tag{5.19}$$

$$\hat{a}^{\sin}(l, k) = \sum_x a(x)\sin(l \cdot x)\sin(k \cdot x) = \frac{1}{2}[\hat{a}(l-k) - \hat{a}(l+k)]. \tag{5.20}$$

We first show that, for all $k, l \in [-\pi, \pi]^d$,

$$-\frac{1}{2}\Delta_k \hat{A}(l) = \frac{1}{2}[\hat{A}(l-k) + \hat{A}(l+k)]\hat{A}(l)[\hat{a}(l) - \hat{a}^{\cos}(l,k)]$$
$$-\hat{A}(l-k)\hat{A}(l)\hat{A}(l+k)\hat{a}^{\sin}(l,k)^2. \tag{5.21}$$

Let $\hat{a}_\pm = \hat{a}(l \pm k)$ and write $\hat{a} = \hat{a}(l)$. Direct computation using (5.13) gives

$$-\frac{1}{2}\Delta_k \hat{A}(l) \tag{5.22}$$

$$= \frac{1}{2}\hat{A}(l)\hat{A}(l+k)\hat{A}(l-k)\Big[[2\hat{a} - \hat{a}_+ - \hat{a}_-] + [2\hat{a}_+\hat{a}_- - \hat{a}\hat{a}_- - \hat{a}\hat{a}_+]\Big]$$

$$= \hat{A}(l)\hat{A}(l+k)\hat{A}(l-k)\Big[[\hat{a}(l) - \hat{a}^{\cos}(l,k)] + [\hat{a}_+\hat{a}_- - \hat{a}(l)\hat{a}^{\cos}(l,k)]\Big],$$

using (5.19) in the last step. By definition, and using the identity $\cos(u+v) = \cos u \cos v - \sin u \sin v$,

$$\hat{a}_+\hat{a}_- = \sum_{x,y} a(x)a(y)\cos((l+k)\cdot x)\cos((l-k)\cdot y)$$

$$= \hat{a}^{\cos}(l,k)^2 - \hat{a}^{\sin}(l,k)^2. \tag{5.23}$$

Substitution of (5.23) in (5.22) gives

$$-\frac{1}{2}\Delta_k \hat{A}(l) = \hat{A}(l-k)\hat{A}(l)\hat{A}(l+k)\big[[\hat{a}(l) - \hat{a}^{\cos}(l,k)][1 - \hat{a}^{\cos}(l,k)] - \hat{a}^{\sin}(l,k)^2\big]. \tag{5.24}$$

Finally, we use (5.19) to rewrite $1 - \hat{a}^{\cos}(l,k)$ and obtain (5.21).

Now we use (5.21) to prove (5.16). First, we note that

$$|\hat{a}(l) - \hat{a}^{\cos}(l,k)| \leq \sum_x [1 - \cos(k \cdot x)]\,|\cos(l \cdot x)|\,|a(x)|$$

$$\leq \hat{a}^{\mathrm{av}}(0) - \hat{a}^{\mathrm{av}}(k). \tag{5.25}$$

Also, by (5.20) and the Cauchy-Schwarz inequality,

$$\hat{a}^{\sin}(k,l)^2 \leq \left(\sum_x |a(x)|\sin^2(k \cdot x)\right)\left(\sum_x |a(x)|\sin^2(l \cdot x)\right). \tag{5.26}$$

With the elementary estimate $\sin^2 t = 1 - \cos^2 t \leq 2[1 - \cos t]$, this gives

$$\hat{a}^{\sin}(k,l)^2 \leq \sum_x |a(x)|[1 - \cos^2(k \cdot x)]\sum_y |a(y)|[1 - \cos^2(l \cdot y)]$$

$$\leq 4[\hat{a}^{\mathrm{av}}(0) - \hat{a}^{\mathrm{av}}(k)][\hat{a}^{\mathrm{av}}(0) - \hat{a}^{\mathrm{av}}(l)]. \tag{5.27}$$

The desired estimate (5.16) then follows from (5.21), (5.25) and (5.27).   ∎

5.2 Convergence of the Expansion

In this section we prove convergence of the lace expansion, assuming (5.2), and also prove Theorem 5.1. Convergence will be proved in the process of proving the following theorem, which shows that if the critical simple random walk bubble diagram is sufficiently small, then the critical self-avoiding walk bubble diagram is also small. (In both diagrams, the trivial term 1 is omitted to obtain a small quantity.)

Theorem 5.8. *There is a $\beta_0 > 0$ and a constant c such that if (5.2) holds with $\beta \leq \beta_0$, then $B(z_c) - 1$ is less than $c\beta$.*

Proof of Theorem 5.1. This is an immediate consequence of Proposition 5.3, Theorem 5.8, and Theorem 2.3. ∎

We will prove Theorem 5.8 in the remainder of Chap. 5. The proof is inspired by the method of [32]. It is possible to go beyond Theorem 5.8 in several respects, and this will be discussed in Chap. 6. In particular, critical exponents of the nearest-neighbour strictly self-avoiding walk in dimensions $d \geq 5$ are computed in [97, 98].

It is not obvious, at first, how to approach the issue of convergence of the lace expansion. The conclusion of Theorem 5.8 is that $\|H_{z_c}\|_2$ is small. On the other hand, recall from (4.11) that

$$\|H_z * G_z\|_\infty \leq \|H_z\|_\infty + \|H_z\|_2^2. \tag{5.28}$$

To use this to perform the sum over N in Theorem 4.1 to estimate Π_z, we already need to know that $\|H_z\|_2^2$ is small uniformly in $z < z_c$. The following elementary lemma will be used to allow us to pick ourselves up by our bootstraps.

Lemma 5.9. *Let $a < b$, let f be a continuous function on the interval $[z_1, z_2)$, and assume that $f(z_1) \leq a$. Suppose for each $z \in (z_1, z_2)$ that if $f(z) \leq b$ then in fact $f(z) \leq a$. Then $f(z) \leq a$ for all $z \in [z_1, z_2)$.*

Proof. By hypothesis, $f(z)$ cannot lie strictly between a and b for any $z \in (z_1, z_2)$. Since $f(z_1) \leq a$, it follows by continuity that $f(z) \leq a$ for all $z \in [z_1, z_2)$. ∎

For $z \in [0, z_c)$, we define $p(z) \in [0, 1/|\Omega|)$ by

$$\hat{G}_z(0) = \chi(z) = \frac{1}{1 - p(z)|\Omega|} = \hat{C}_{p(z)}(0), \tag{5.29}$$

which is equivalent to

$$p(z)|\Omega| = 1 - \frac{1}{\chi(z)}. \tag{5.30}$$

Our choice of f is motivated, in part, by the intuition that $\hat{G}_z(k)$ and $\hat{C}_{p(z)}(k)$ are comparable in size. We also expect $\frac{1}{2}\Delta_k\hat{G}_z(l)$ and $\frac{1}{2}\Delta_k\hat{C}_{p(z)}(l)$ to be comparable. However, rather than comparing the latter directly, we will compare $\frac{1}{2}\Delta_k\hat{G}_z(l)$ with

$$U_{p(z)}(k,l) = 16\hat{C}_{p(z)}(k)^{-1}\left(\hat{C}_{p(z)}(l-k)\hat{C}_{p(z)}(l) + \hat{C}_{p(z)}(l+k)\hat{C}_{p(z)}(l)\right.$$
$$\left. +\hat{C}_{p(z)}(l-k)\hat{C}_{p(z)}(l+k)\right), \qquad (5.31)$$

which can be seen using (5.16) to be an upper bound for $\frac{1}{2}|\Delta_k\hat{C}_{p(z)}(l)|$.

We will apply Lemma 5.9 with $z_1 = 0$, $z_2 = z_c$, $b = 4$, $a = 1 + \text{const}\beta$ (the constant being determined in the course of the proof), and

$$f(z) = \max\{f_1(z), f_2(z), f_3(z)\}, \qquad (5.32)$$

where

$$f_1(z) = z|\Omega|, \quad f_2(z) = \sup_{k\in[-\pi,\pi]^d}\frac{|\hat{G}_z(k)|}{\hat{C}_{p(z)}(k)}, \qquad (5.33)$$

$$f_3(z) = \sup_{k,l\in[-\pi,\pi]^d}\frac{\frac{1}{2}|\Delta_k\hat{G}_z(l)|}{U_{p(z)}(k,l)}. \qquad (5.34)$$

Note that the factor $\hat{C}_{p(z)}(k)^{-1}$ in the denominator of f_3 becomes arbitrarily small when $k = 0$ and $z \to z_c^-$. We will verify in Lemmas 5.12, 5.14 and 5.16 that the hypotheses of Lemma 5.9 hold when β is sufficiently small. From this, we can conclude that $f(z) \le a = 1 + \text{const}\beta$ uniformly in $z \in [0, z_c)$.

Proof of Theorem 5.8. We will show below in Lemma 5.10 that it follows from $f(z) \le a$ (which we will conclude as noted above) that $\|H_z\|_2^2 \le c_a\beta$, where c_a is the constant of Lemma 5.10 when $f(z) \le K = a$. This proves $\|H_z\|_2^2 \le c_a\beta$ uniformly in $z < z_c$. By the monotone convergence theorem, this implies that

$$\|H_{z_c}\|_2^2 = \lim_{z\to z_c^-}\|H_z\|_2^2 \le c_a\beta, \qquad (5.35)$$

which proves Theorem 5.8 since $B(z_c) - 1 = \|H_{z_c}\|_2^2$ by (4.12). ∎

Note that the inequality $f_2(z) \le a$ implies the *infrared bound*

$$\hat{G}_z(k) \le a\hat{C}_{p(z)}(k) \qquad (5.36)$$

(we will actually prove in (5.53)–(5.60) that $\hat{G}_z(k)/\hat{C}_{p(z)}(k) = 1+O(\beta)$, which implies, in particular, that $\hat{G}_z(k) \ge 0$, permitting removal of the absolute value on the left hand side of (5.36)).

Before going into the details, the basic strategy is as follows. First, it is straightforward to verify the two hypotheses on f in Lemma 5.9 that f is continuous and that $f(0) \le a$, and the main work goes into verifying that $f(z) \le b$ implies that $f(z) \le a$. For this, we use the assumption $f(z) \le b$ to compare norms of H_z with norms of $C_{p(z)}$, and use (5.2) and Lemma 5.5 to see that the latter are small. We then apply Theorem 4.1 to conclude that $\hat{\Pi}_z(k)$ is as small as we like, assuming that β is sufficiently small. Importantly, this can be done even for a poor (large) value of b, because the effect of taking β small compensates for the lack of sharpness in the bound $f(z) \leq b$. This implies that $G_z(k)$ is close to a simple random walk quantity, and from this we will be able to conclude the sharper bound $f(z) \leq a$. The details are carried out below.

Lemma 5.10. *Fix $z \in (0, z_c)$, assume that f of (5.32) obeys $f(z) \leq K$, and assume (5.2). Then there is a constant c_K, independent of z, such that*

$$\|[1 - \cos(k \cdot x)]H_z\|_\infty \leq c_K(1 + \beta)\hat{C}_{p(z)}(k)^{-1}, \qquad (5.37)$$

$$\|H_z\|_2^2 \leq c_K\beta, \qquad \|H_z\|_\infty \leq c_K\beta. \qquad (5.38)$$

Proof. As in (5.14),

$$\|[1 - \cos(k \cdot x)]H_z\|_\infty = \|[1 - \cos(k \cdot x)]G_z\|_\infty$$
$$\leq \frac{1}{2}\|\Delta_k \hat{G}_z\|_1. \qquad (5.39)$$

It then follows from $f_3(z) \leq K$, the Cauchy–Schwarz inequality, and (5.8) that

$$\|[1 - \cos(k \cdot x)]H_z\|_\infty \leq 16K\hat{C}_{p(z)}(k)^{-1}3\|\hat{C}_{p(z)}\|_2^2$$
$$\leq 48(1 + 3\beta)K\hat{C}_{p(z)}(k)^{-1}, \qquad (5.40)$$

which gives (5.37).

Next, we estimate $\|H_z\|_2^2$. We first use subadditivity and $f_1(z) \leq K$ to obtain

$$H_z(x) \leq z|\Omega|(D * G_z)(x) \leq K(D * G_z)(x). \qquad (5.41)$$

Using $f_2(z) \leq K$, the Parseval relation, and (5.2), this implies that

$$\|H_z\|_2^2 \leq K^2\|D * G_z\|_2^2 = K^2\|\hat{D}\hat{G}_z\|_2^2$$
$$\leq K^4\|\hat{D}\hat{C}_{p(z)}\|_2^2 = K^4\|D * C_{p(z)}\|_2^2$$
$$\leq K^4\|D * C_{1/|\Omega|}\|_2^2 = K^4\|\hat{D}[1 - \hat{D}]^{-1}\|_2^2 \leq K^4\beta. \qquad (5.42)$$

This proves the first bound of (5.38).

Iteration of (5.41) gives $H_z(x) \leq KD(x) + K^2(D * D * G_z)(x)$. Therefore,

$$\begin{aligned}
\|H_z\|_\infty &\leq K\|D\|_\infty + K^2\|\hat{D}^2\hat{G}_z\|_1 \\
&\leq K\beta + K^3\|\hat{D}^2\hat{C}_{p(z)}\|_1 \\
&= K\beta + K^3(D * D * C_{p(z)})(0) \\
&\leq K\beta + K^3(D * D * C_{p(z)} * C_{p(z)})(0) \\
&\leq K\beta + K^3(D * D * C_{1/|\Omega|} * C_{1/|\Omega|})(0) \\
&\leq K\beta + K^3\beta,
\end{aligned} \tag{5.43}$$

using (5.7) in the second inequality, and the inverse Fourier transform and (5.2) in the last.   ∎

Remark. The bounds of Lemma 5.10 can be combined with Theorem 4.1 to give bounds on $\Pi^{(N)}$, and hence on Π. This is the content of the following lemma. Note that once we have verified that the hypotheses of Lemma 5.9 all hold, we can conclude that $f(z) \leq a = 1 + \text{const}\beta$, thereby verifying the hypothesis $f(z) \leq K$ of Lemma 5.11 with $K = a$. After the fact, this then gives unconditional bounds on Π_z (of course assuming (5.2)). These bounds are then of lasting importance (see, e.g., Exercises 5.17–5.18).

Lemma 5.11. *Fix $z \in (0, z_c)$, assume that f of (5.32) obeys $f(z) \leq K$, and assume that (5.2) holds. There is a constant $\bar{c}_K$, independent of z, such that if β is sufficiently small (independent of z), then*

$$\sum_{x \in \mathbb{Z}^d} |\Pi_z(x)| \leq \bar{c}_K\beta, \tag{5.44}$$

$$\sum_{x \in \mathbb{Z}^d} [1 - \cos(k \cdot x)]|\Pi_z(x)| \leq \bar{c}_K\beta\hat{C}_{p(z)}(k)^{-1}. \tag{5.45}$$

Proof. It follows from Theorem 4.1, Lemma 5.10, and the estimate (5.28) that there is a constant c'_K such that

$$\sum_{x \in \mathbb{Z}^d} \Pi_z^{(N)}(x) \leq (c'_K\beta)^N \tag{5.46}$$

for all $N \geq 1$, and

$$\sum_{x \in \mathbb{Z}^d} [1 - \cos(k \cdot x)]\Pi_z^{(N)}(x) \begin{cases} = 0 & \text{if } N = 1 \\ \leq \hat{C}_{p(z)}(k)^{-1}N^2(c'_K\beta)^{N-1} & \text{if } N \geq 2. \end{cases} \tag{5.47}$$

The bounds (5.44)–(5.45) then follow immediately.   ∎

We now confirm that f of (5.32) obeys the hypotheses of Lemma 5.9, with $z_1 = 0$, $z_2 = z_c$, $b = 4$ and $a = 1 + \text{const}\beta$ (the particular value 4 for b is not an essential choice). We first verify that $f(0) = 1$, which is of course less than a.

Lemma 5.12. *The function f of (5.32) obeys $f(0) = 1$.*

Proof. By definition, $f_1(0) = 0$. Also, $p(0) = 0$ by (5.30) and hence $f_2(0) = 1$. Finally, $f_3(0) = 0$. ∎

To prove the continuity of f, we will use the following elementary lemma.

Lemma 5.13. *Let $(f_\alpha)_{\alpha \in A}$ be an equicontinuous family of functions on an interval $[t_1, t_2]$, and suppose that $\sup_{\alpha \in A} f_\alpha(t) < \infty$ for each $t \in [t_1, t_2]$. Then $\sup_{\alpha \in A} f_\alpha$ is continuous on $[t_1, t_2]$.*

Proof. Let $\bar{f} = \sup_{\alpha \in A} f_\alpha$, and let $\epsilon > 0$ be given. The statement that $(f_\alpha)_{\alpha \in A}$ is equicontinuous means that there is a $\delta > 0$ such that $|f_\alpha(s) - f_\alpha(t)| < \epsilon/2$ whenever $|s - t| < \delta$, uniformly in $\alpha \in A$. Fix s, t with $|s - t| < \delta$, and assume without loss of generality that $\bar{f}(s) \geq \bar{f}(t)$. Choose α' such that $0 \leq \bar{f}(s) - f_{\alpha'}(s) < \epsilon/2$. Then

$$
\begin{aligned}
0 \leq \bar{f}(s) - \bar{f}(t) &\leq \bar{f}(s) - f_{\alpha'}(t) \\
&\leq [\bar{f}(s) - f_{\alpha'}(s)] + |f_{\alpha'}(s) - f_{\alpha'}(t)| \\
&< \frac{\epsilon}{2} + \frac{\epsilon}{2} = \epsilon.
\end{aligned}
\tag{5.48}
$$

Therefore, $\bar{f}$ is continuous. ∎

Lemma 5.14. *The function f of (5.32) is continuous on the interval $[0, z_c)$.*

Proof. It suffices to show that each of f_1, f_2, f_3 is continuous on $[0, z_c)$. The function f_1 is linear, so it is certainly continuous.

For f_2, it suffices to show that f_2 is continuous in $[0, r]$ for every $r < z_c$. By Lemma 5.13, it suffices to show that $|\hat{G}_z(k)|/\hat{C}_{p(z)}(k)$ is equicontinuous in $z \in [0, r]$. Here α is k. Since $(|f_\alpha|)_{\alpha \in A}$ is an equicontinuous family whenever $(f_\alpha)_{\alpha \in A}$ is, it suffices to obtain a bound on the derivative

$$
\frac{d}{dz} \frac{\hat{G}_z(k)}{\hat{C}_{p(z)}(k)} = \frac{1}{\hat{C}_{p(z)}(k)^2} \left[\hat{C}_{p(z)}(k) \frac{d\hat{G}_z(k)}{dz} - \hat{G}_z(k) \frac{d\hat{C}_p(k)}{dp} \Big|_{p=p(z)} \frac{dp(z)}{dz} \right],
\tag{5.49}
$$

uniformly in k and in $z \in [0, r]$. This follows from the bounds

$$
\frac{1}{2} \leq \frac{1}{1 - p(z)|\Omega|\hat{D}(k)} = \hat{C}_{p(z)}(k) \leq \hat{C}_{p(z)}(0) = \chi(z) \leq \chi(r),
\tag{5.50}
$$

$|\hat{G}_z(k)| \leq \chi(r)$, $|\frac{d\hat{G}_z(k)}{dz}| \leq \chi'(r)$, $|\frac{d\hat{C}_p(k)}{dp}| \leq |\Omega|\chi(r)^2$, and $\frac{dp(z)}{dz} \leq |\Omega|^{-1}\chi'(r)$.

For f_3, it again suffices to show continuity in $[0, r]$ for every $r < z_c$. Again we show equicontinuity on $[0, r]$ for every $r < z_c$, by obtaining a uniform bound on the derivative with respect to z, and this follows as before. ∎

Exercise 5.15. Fill in the missing details in the continuity proof of f_3.

Finally, we verify that f obeys the substantial hypothesis of Lemma 5.9.

Lemma 5.16. *Fix $z \in (0, z_c)$ and suppose that $f(z) \leq 4$. If (5.2) holds with β sufficiently small (independent of z), then it is in fact the case that $f(z) \leq 1 + c\beta$ for some $c > 0$ independent of z.*

Proof. For $f_1(z)$, we simply note that $\chi(z) > 0$ and hence, by (3.30),

$$\chi(z)^{-1} = 1 - z|\Omega| - \hat{\Pi}_z(0) > 0. \tag{5.51}$$

Therefore, by Lemma 5.11,

$$f_1(z) = z|\Omega| < 1 - \hat{\Pi}_z(0) \leq 1 + \bar{c}_4\beta \tag{5.52}$$

if β is sufficiently small.

For f_2, we first write $\hat{F}_z(k) = 1/\hat{G}_z(k)$, so that

$$\frac{\hat{G}_z(k)}{\hat{C}_{p(z)}(k)} = \frac{1 - p(z)|\Omega|\hat{D}(k)}{\hat{F}_z(k)} = 1 + \frac{1 - p(z)|\Omega|\hat{D}(k) - \hat{F}_z(k)}{\hat{F}_z(k)}. \tag{5.53}$$

We will show that the last term on the right hand side is $O(\beta)$, which implies that $f_2(z) = 1 + O(\beta)$.

We first obtain bounds on the numerator of the last term in (5.53), and afterwards consider the denominator. By (5.30) and (3.30), $p(z)|\Omega| = 1 - \hat{F}_z(0) = z|\Omega| + \hat{\Pi}_z(0)$, and thus the numerator of the last term in (5.53) is

$$1 - p(z)|\Omega|\hat{D}(k) - \hat{F}_z(k) = \hat{\Pi}_z(0)[1 - \hat{D}(k)] - [\hat{\Pi}_z(0) - \hat{\Pi}_z(k)]. \tag{5.54}$$

This is bounded above by $4\bar{c}_4\beta$, by (5.44). Additionally, by (5.44)–(5.45), it is also bounded above by

$$\bar{c}_4\beta[1 - \hat{D}(k)] + \bar{c}_4\beta[1 - p(z)|\Omega|\hat{D}(k)]. \tag{5.55}$$

Since

$$[1 - \hat{D}(k)]\hat{C}_{p(z)}(k) = 1 + \hat{D}(k)\frac{p(z)|\Omega| - 1}{1 - p(z)|\Omega|\hat{D}(k)} \leq 2, \tag{5.56}$$

the numerator of (5.53) is bounded by

$$3\bar{c}_4\beta[1 - p(z)|\Omega|\hat{D}(k)] \leq 3\bar{c}_4\beta\left[\hat{F}_z(0) + [1 - \hat{D}(k)]\right]. \tag{5.57}$$

The denominator of (5.53) is

$$\begin{aligned}\hat{F}_z(k) &= \hat{F}_z(0) + [\hat{F}_z(k) - \hat{F}_z(0)] \\ &= \hat{F}_z(0) + z|\Omega|[1 - \hat{D}(k)] + [\hat{\Pi}_z(0) - \hat{\Pi}_z(k)].\end{aligned} \tag{5.58}$$

For $z \leq 1/2|\Omega|$, we use $\hat{F}_z(0) \geq \hat{C}_z(0)^{-1} \geq \frac{1}{2}$, $1 - \hat{D}(k) \geq 0$, and (5.44) to see that

$$\hat{F}_z(k) \geq \hat{F}_z(0) - 2\bar{c}_4\beta \geq \frac{1}{2} - 2\bar{c}_4\beta. \tag{5.59}$$

For $1/2|\Omega| \leq z < z_c$, we use $\hat{F}_z(0) > 0$, (5.45) and $1 - p(z)|\Omega|\hat{D}(k) = 1 - (1 - \hat{F}_z(0))\hat{D}(k) \leq 1 - \hat{D}(k) + \hat{F}_z(0)$ to obtain

$$\hat{F}_z(k) \geq \hat{F}_z(0) + \frac{1}{2}[1 - \hat{D}(k)] - \bar{c}_4\beta[1 - p(z)|\Omega|\hat{D}(k)]$$

$$\geq \left[\frac{1}{2} - \bar{c}_4\beta\right]\left[\hat{F}_z(0) + [1 - \hat{D}(k)]\right]. \tag{5.60}$$

In either case, combining these inequalities with the bounds obtained above for the numerator of (5.53) gives $f_2(z) = 1 + O(\beta)$.

Finally, we consider f_3. We write

$$\hat{g}_z(k) = z|\Omega|\hat{D}(k) + \hat{\Pi}_z(k), \tag{5.61}$$

so that

$$\hat{G}_z(k) = \frac{1}{1 - \hat{g}_z(k)}. \tag{5.62}$$

Note that $g_z(x) = g_z(-x)$, so we can apply Lemma 5.7 to obtain

$$\frac{1}{2}|\Delta_k\hat{G}_z(l)| \leq \frac{1}{2}[\hat{G}_z(l - k) + \hat{G}_z(l + k)]\hat{G}_z(l)[\hat{g}_z^{\mathrm{av}}(0) - \hat{g}_z^{\mathrm{av}}(k)] \tag{5.63}$$

$$+ 4\hat{G}_z(l - k)\hat{G}_z(l)\hat{G}_z(l + k)[\hat{g}_z^{\mathrm{av}}(0) - \hat{g}_z^{\mathrm{av}}(k)][\hat{g}_z^{\mathrm{av}}(0) - \hat{g}_z^{\mathrm{av}}(l)].$$

Using $f_2(z) \leq 1 + O(\beta)$, we can bound each factor of $\hat{G}_z$ above by $[1 + O(\beta)]\hat{C}_{p(z)}$. Also,

$$\hat{g}_z^{\mathrm{av}}(0) - \hat{g}_z^{\mathrm{av}}(k) \leq \sum_x [1 - \cos(k \cdot x)][z|\Omega|D(x) + |\Pi_z(x)|]$$

$$\leq z|\Omega|[1 - \hat{D}(k)] + \bar{c}_4\beta\hat{C}_{p(z)}(k)^{-1}$$

$$\leq [2 + O(\beta)]\hat{C}_{p(z)}(k)^{-1}, \tag{5.64}$$

using (5.45) for the second inequality, and $f_1(z) \leq 1 + O(\beta)$ and (5.56) for the third. Combining these bounds gives $f_3(z) \leq 1 + O(\beta)$.

This completes the proof that $f(z) \leq 1 + O(\beta)$. ∎

This completes the proof that f obeys the hypotheses of Lemma 5.9, and also completes the proof of Theorem 5.8.

Exercise 5.17. (a) Give a monotonicity argument to conclude that the factor $\hat{C}_{p(z)}(k)^{-1}$ in (5.45) can be replaced by $1 - \hat{D}(k)$.
(b) Use the result of (a) to prove that $\sum_x |x|^2|\Pi_z(x)|$ is bounded above by $O(\beta\sigma^2)$ uniformly in $z < z_c$, where $\sigma^2 = \sum_x |x|^2 D(x)$.
(c) The *correlation length of order 2*, $\xi_2(z)$, is defined by

$$\xi_2(z)^2 = \frac{1}{\chi(z)} \sum_x |x|^2 G_z(x). \tag{5.65}$$

Prove that $\xi_2(z) \simeq (1 - z/z_c)^{-1/2}$.

As a final observation, we note that by (5.46) and dominated convergence, $\hat{\Pi}_{z_c}(k)$ is finite and is equal to the limit of $\hat{\Pi}_z(k)$ as z approaches z_c from the left. Since $\chi(z)$ diverges to infinity in this limit, it follows from (3.30) that

$$1 - z_c|\Omega| - \hat{\Pi}_{z_c}(0) = 0, \tag{5.66}$$

and hence

$$z_c = \frac{1}{|\Omega|} \left(1 - \hat{\Pi}_{z_c}(0) \right) = \frac{1}{|\Omega|} + O\Big(\frac{1}{|\Omega|^2}\Big), \tag{5.67}$$

where we have used $|\hat{\Pi}_{z_c}(0)| \leq O(\beta) = O(|\Omega|^{-1})$ (and we assume $d > 4$ for the spread-out model). For the spread-out model in dimensions $d > 4$, an extension of (5.67) can be found in [118].[1] See [176] for related results for the spread-out model in dimensions $d \leq 4$. For the nearest-neighbour model, (5.67) is the first step in the proof of the asymptotic formula (2.8) for $\mu = 1/z_c$. The following exercise pushes (5.67) a bit further.

Exercise 5.18. Consider the nearest-neighbour model.
(a) Fix an integer $m \geq 1$. Show that $\|[1 - \hat{D}]^{-m}\|_1$ is nonincreasing in $d > 2m$. Hint: $A^{-m} = \Gamma(m)^{-1} \int_0^\infty u^{m-1} e^{-uA} du$.
(b) Let $H_z^{(j)}(x) = \sum_{m=j}^\infty c_m(x) z^m$. Show that $\|H_{z_c}^{(j)}\|_\infty \leq O(d^{-j/2})$, where the constant may depend on j. To do so, it is helpful first to show that $\|\hat{D}^{2j}\|_1 \leq c_j (2d)^{-j}$ for some constant c_j depending on $j = 1, 2, \ldots$.
(c) Prove that

$$\hat{\Pi}_{z_c}^{(1)}(0) = \frac{1}{2d} + \frac{3}{(2d)^2} + O\Big(\frac{1}{(2d)^3}\Big). \tag{5.68}$$

(d) Prove that

$$\hat{\Pi}_{z_c}^{(2)}(0) = \frac{1}{(2d)^2} + O\Big(\frac{1}{(2d)^3}\Big). \tag{5.69}$$

(e) Conclude that the connective constant $\mu = z_c^{-1}$ obeys

$$\mu = 2d - 1 - \frac{1}{2d} + O\Big(\frac{1}{(2d)^2}\Big). \tag{5.70}$$

The strategy in this exercise is based on that used in [53, 122, 123], and is simpler than that used in [101]. Equation (5.70) was first proved in [140], using completely different methods.

---

[1] The results of [118] are expressed in terms of p_c defined by $p_c = z_c|\Omega|$.

5.3 Finite Bubble vs Small Bubble

According to Theorem 2.3, the susceptibility obeys the mean-field behaviour $\chi(z) \simeq (1 - z/z_c)^{-1}$ if the critical bubble diagram $B(z_c)$ is finite. On the other hand, convergence of the lace expansion has been proved only when $B(z_c) - 1$ is *small*. This leads to the restrictions that the dimension be large for the nearest-neighbour model, or that L be large for the spread-out model in dimensions $d > 4$, in our use of Proposition 5.3 to drive the convergence proof.

For the nearest-neighbour model in dimension $d = 5$, it was shown in [97,98] that $B(z_c) - 1 \leq 0.493$. This is small, although not very small. With considerable effort, and with a computer-assisted proof, convergence of the lace expansion was proved in [97,98] for the nearest-neighbour model in dimensions $d \geq 5$.

It would be of great interest to find a proof of the bubble condition that would be applicable in situations where the bubble diagram could be large, rather than relying on it being small.

5.4 Differential Equality and the Bubble Condition

It is instructive now to revisit Theorem 2.3, which used inclusion-exclusion to give upper and lower bounds on the derivative of $z\chi(z)$. As in the proof of Lemma 5.16, we write

$$\hat{F}_z(0) = \frac{1}{\chi(z)} = 1 - z|\Omega| - \hat{\Pi}_z(0). \tag{5.71}$$

Then direct calculation gives

$$\frac{\mathrm{d}[z\chi(z)]}{\mathrm{d}z} = \left(\hat{F}_z(0) - z\frac{\mathrm{d}\hat{F}_z(0)}{\mathrm{d}z} \right) \frac{1}{\hat{F}_z(0)^2} = V(z)\chi(z)^2, \tag{5.72}$$

with

$$V(z) = 1 + z\frac{\mathrm{d}\hat{\Pi}_z(0)}{\mathrm{d}z} - \hat{\Pi}_z(0). \tag{5.73}$$

The identity (5.72) gives an identity in place of the inequalities of (2.35), and corresponds to inclusion-exclusion carried out to all orders.

It is significant that $V(z_c)$ is finite, under the basic assumption of Chap. 5 that (5.2) is sufficiently small. We have already seen in Sect. 5.2 that $\hat{\Pi}_{z_c}(0)$ is finite. To see that $V(z_c)$ is finite, we must verify that the derivative $\mathrm{d}\hat{\Pi}_{z_c}(0)/\mathrm{d}z$ is also finite. Here is a sketch of a proof of this last fact.

It suffices to obtain a bound on

$$\sum_{m=1}^{\infty} m\hat{\pi}_m^{(N)}(0)z_c^{m-1} \tag{5.74}$$

which is summable in N. As in the proof of (4.10), we associate to $\hat{\pi}_m^{(N)}(0)$ a diagram consisting of $2N - 1$ subwalks, whose lengths $m_1, \ldots, m_{2N-1}$ sum to m. We decompose the factor m in (5.74) as $m = \sum_{j=1}^{2N-1} m_j$ and obtain a sum of $2N - 1$ terms. In the jth term, there is a factor m_j associated to the jth line. We apply Lemma 4.6 to estimate the jth term, associating the infinity norm to the special line. Then we use the bound

$$\| \sum_{m=1}^{\infty} m c_m(x) z_c^{m-1} \|_\infty = \| \mathrm{d} H_{z_c}(x)/\mathrm{d}z \|_\infty$$

$$\leq \| H_{z_c} * G_{z_c} \|_\infty \leq \| H_{z_c} \|_\infty + \| H_{z_c} \|_2^2, \quad (5.75)$$

and draw the desired conclusion. The first inequality of (5.75) follows as in the upper bound of (2.35), and the second inequality is (4.11).

This shows that, as $z \to z_c^-$,

$$\frac{\mathrm{d}[z\chi(z)]}{\mathrm{d}z} \sim V(z_c)\chi(z)^2. \quad (5.76)$$

The left hand side is equal to the generating function of two mutually-avoiding self-avoiding walks starting from the origin. The asymptotic formula (5.76) shows that this generating function behaves in the same way as the generating function for two *independent* self-avoiding walks, up to a vertex factor $V(z_c)$ which takes into account the local effect of the mutual avoidance.

Exercise 5.19. Prove that when β of (5.2) is sufficiently small, the susceptibility obeys the asymptotic formula

$$\chi(z) \sim A(1 - z/z_c)^{-1} \text{ as } z \to z_c^-, \quad (5.77)$$

with $A = z_c^{-1}[|\Omega| + \frac{\mathrm{d}}{\mathrm{d}z}\hat{\Pi}_{z_c}(0)]^{-1}$. This improves the conclusion of Theorem 5.1 to an asymptotic formula, and also avoids any appeal to Theorem 2.3.

Further Results for the Self-Avoiding Walk

The proof of convergence of the lace expansion was presented in Chap. 5 in the context of proving the bubble condition, which is the simplest meaningful result that can be derived from convergence. However, it is possible to go substantially further, and in Sect. 6.1 we discuss several extensions for the self-avoiding walk in dimensions $d > 4$. In Sect. 6.2, we discuss a result of van der Hofstad [108] for a 1-dimensional self-avoiding walk. In Sect. 6.3, we discuss a result of Ueltschi [204] for a self-avoiding walk with nearest-neighbour attraction. Finally, in Sect. 6.4, we discuss applications of the lace expansion to networks of mutually-avoiding self-avoiding walks in dimensions $d > 4$.

6.1 The Self-Avoiding Walk in Dimensions $d > 4$

The lace expansion was invented by Brydges and Spencer [45] to prove the following theorem for the weakly self-avoiding walk (see (2.2) and (2.14)).

Theorem 6.1. *Let* $d > 4$ *and consider the nearest-neighbour weakly self-avoiding walk. There is a* $\lambda_0 > 0$ *and an* $\epsilon > 0$ *such that for all* $0 < \lambda \le \lambda_0$ *there are constants* A_λ *and* v_λ *such that for every* $k \in \mathbb{R}^d$,

$$\hat{c}_n^{(\lambda)}(k/\sqrt{v_\lambda n}) = A_\lambda \mu_\lambda^n e^{-k^2/2d}[1 + O(n^{-\epsilon})], \tag{6.1}$$

$$\mathbb{E}_n^{(\lambda)}|\omega(n)|^2 = v_\lambda n[1 + O(n^{-\epsilon})]. \tag{6.2}$$

The above result was revisited in [81, 144], where the lace expansion was treated as an example of a cluster expansion in statistical mechanics. The fact that this is a natural setting for the lace expansion has been emphasized in [40, 209].

A version of Theorem 6.1 was also obtained in [113], using a different approach to the lace expansion based on induction. The inductive approach notes that in the recursion

$$\hat{c}_n(k) = |\Omega|\hat{D}(k)\hat{c}_{n-1}(k) + \sum_{m=1}^{n} \hat{\pi}_m(k)\hat{c}_{n-m}(k) \tag{6.3}$$

of (3.26), the right hand side can be analyzed using only information about $\hat{c}_j(k)$ for $j < n$. In fact, this is clear for those $\hat{c}_j(k)$ occurring explicitly on the right hand side, and diagrammatic estimates can be used to bound $\hat{\pi}_m(k)$ using only $\hat{c}_j(k)$ for $j < m$. The results of [113] gives error estimates for $\hat{c}_n^{(\lambda)}(0)$ and the mean-square displacement that are likely optimal, namely

$$\hat{c}_n^{(\lambda)}(0) = A_\lambda \mu_\lambda^n [1 + O(n^{-(d-4)/2})], \tag{6.4}$$

$$\mathbb{E}_n^{(\lambda)} |\omega(n)|^2 = \begin{cases} v_\lambda n[1 + O(n^{-1 \wedge (d-4)/2})] & (d \neq 5) \\ v_\lambda n[1 + O(n^{-1} \log n)] & (d = 5). \end{cases} \tag{6.5}$$

In particular, the power in the error term for $\hat{c}_n^{(\lambda)}(0)$ goes to infinity with the dimension.

In [120], an inductive analysis was used to show that under appropriate hypotheses, more general recursion relations have solutions with Gaussian asymptotics. The result of the inductive analysis is quite general, and has been applied in several contexts [119, 121, 124, 125, 204]. In particular, it was used in [124] to prove a version of Theorem 6.1 for the spread-out model (with L of (1.2) sufficiently large) of strictly self-avoiding walks ($\lambda = 1$) in dimensions $d > 4$. Such a result had been obtained previously in [158], using generating function methods rather than induction.

A variant of Theorem 6.1 was proved in [207] for a weakly self-avoiding version of a walk that takes steps of length r parallel to the coordinate axes with probability proportional to r^{-2}. In this case, it was shown that for $d > 2$ and for sufficiently small λ there is convergence to the Cauchy distribution rather than to a Gaussian. A related result was obtained for a spread-out model in [52]. The fact that the upper critical dimension depends on the range of interaction was discussed in [10] for spin systems.

In the series of papers [183–185], it was shown that the conclusions of Theorem 6.1 hold for the nearest-neighbour strictly self-avoiding walk ($\lambda = 1$) if $d \geq d_0$, provided d_0 is sufficiently large. It was also proved that the scaling limit is Brownian motion in this case. This was improved in [97, 98] to give the following theorem (see [97] for additional statements). The proof of the theorem was computer assisted, and used precise numerical estimation of error terms to show that $d = 5$ is large enough (barely!) to give convergence.

Theorem 6.2. *Let $d \geq 5$ and $\lambda = 1$ (strictly self-avoiding walk) and consider the nearest-neighbour model. There are constants A and v (depending on d) such that*

$$c_n = A\mu^n [1 + O(n^{-\epsilon})] \quad \text{for any } \epsilon < \tfrac{d-4}{2} \wedge 1, \tag{6.6}$$

$$\mathbb{E}_n |\omega(n)|^2 = vn[1 + O(n^{-\epsilon})] \quad \text{for any } \epsilon < \tfrac{d-4}{4} \wedge 1. \tag{6.7}$$

The scaling limit is Brownian motion, in the sense that if we linearly interpolate $(vn)^{-1/2} \omega$ to obtain a continuous function from $[0,1]$ to $\mathbb{R}^d$, then this converges weakly to Brownian motion. For $d = 5$, $1 \leq A \leq 1.493$ and $1.098 \leq v \leq 1.803$.

In [173], the diffusion constant v is estimated to take the value $1.4767(13)$ for $d = 5$.

The proofs of the above theorems are based on analysis of Fourier transforms, and they have the character of the central limit theorem. However, they do not provide a *local* central limit theorem, which is a statement that $c_n^{(\lambda)}(x)/c_n^{(\lambda)}$ converges to a suitably normalized[1] version of the Gaussian density

$$p_n(x) = \frac{d^{d/2}}{(2\pi n)^{d/2}} e^{-d|x|^2/2n}, \tag{6.8}$$

directly in x-space. A result in this direction was obtained by Bolthausen and Ritzmann [29], as stated in the following theorem. The method of [29] introduced a new approach to the analysis of the lace expansion, based on fixed-point methods.

Theorem 6.3. *Let $d > 4$ and consider the nearest-neighbour model. There are $\lambda_0 > 0$, $K > 0$ and $\epsilon > 0$ such that for all $0 < \lambda \leq \lambda_0$ there is a v (depending on λ, d) such that for all n and $\|x\|_1$ of the same parity,*

$$\left| \frac{c_n^{(\lambda)}(x)}{c_n^{(\lambda)}} - 2p_{vn}(x) \right| \leq K \left[\frac{1}{\sqrt{n}} p_{vn}(x) + \frac{1}{n^{d/2}} \sum_{j=1}^{n/2} j p_{vj}(x) \right]. \tag{6.9}$$

The factor 2 on the left hand side of (6.9) is there for the same parity reason that it is present for simple random walk (see, e.g., [149]). Namely, if we sum the first term on the left hand side over x having the same parity as n, we get 1, but we only get $\frac{1}{2}$ (approximately) if we sum $p_{vn}(x)$ over x with fixed parity.

Exercise 6.4. The right hand side of (6.9) is somewhat opaque, and this exercise considers two consequences of (6.9).
(a) If we sum (6.9) over n, the Gaussian leading term $\sum_{n=0}^{\infty} 2p_{vn}(x)$ behaves like a multiple of $|x|^{2-d}$. Summation of the error term gives as upper bound a multiple of

$$\sum_{j=1}^{\infty} \left(\frac{1}{j^{(d+1)/2}} + \frac{1}{j^{d-2}} \right) e^{-d|x|^2/2vj}. \tag{6.10}$$

By approximating the sum by an integral, this can be seen to be $O(|x|^{1-d} + |x|^{6-2d})$. This is smaller than the leading term $|x|^{2-d}$, and thus provides a proof that $\eta = 0$ (see (2.27)) under the assumptions of Theorem 6.3. Fill in the details of this discussion.
(b) Show that (6.9) implies that $c_n^{(\lambda)}(x)/c_n^{(\lambda)}$ is bounded above by a multiple of $n^{-d/2}$.

In [120, 124], the following averaged version of a local central limit theorem is proved. For the averaging, we denote the cube of radius R centred

---

[1] The function $p_n(x)$ of (6.8) is such that $\int_{\mathbb{R}^d} |x|^2 p_n(x) d^d x = n$ in all dimensions.

at $x \in \mathbb{Z}^d$ by

$$C_R(x) = \{y \in \mathbb{Z}^d : \|x - y\|_\infty \leq R\}. \tag{6.11}$$

We use $\lfloor x \rfloor$ to denote the closest lattice point in $\mathbb{Z}^d$ to $x \in \mathbb{R}^d$ (with an arbitrary rule to break ties).

Theorem 6.5. *Consider the spread-out model of strictly self-avoiding walks. Let $d > 4$. There is an $L_0 = L_0(d)$ such that for $L \geq L_0$, the following holds. Let R_n be any sequence with $\lim_{n\to\infty} R_n = \infty$ and $\lim_{n\to\infty} R_n n^{-1/2} = 0$. Then for all $x \in \mathbb{R}^d$ with $x^2 [\log R_n]^{-1}$ sufficiently small, as $n \to \infty$,*

$$\frac{1}{(2R_n + 1)^d} \sum_{y \in C_{R_n}(\lfloor x\sqrt{vn}\rfloor)} c_n(y) = A\mu^n \left(\frac{d}{2\pi n v}\right)^{d/2} e^{-dx^2/2}[1+o(1)]. \tag{6.12}$$

Note that (6.12) is not an immediate consequence of the convergence of the Fourier transform of $c_n(x)$ to a Gaussian limit, which implies instead that sums over sets of volume $n^{d/2}$ converge to integrals of the Gaussian density over the scaled set. The arbitrarily slow growth of R_n in Theorem 6.5 probes $c_n(x)$ on a smaller scale.

See also Theorem 6.9 below for a related result, also proved using the inductive method of [120], which gives upper and lower bounds on $\sup_x c_n(x)$.

Alternate proofs of convergence of the lace expansion, which avoid use of Fourier transforms, were developed in [90, 91], where the following theorem was proved. Theorem 6.6 provides a statement that $\eta = 0$ for the self-avoiding walk in dimensions $d > 4$.

Theorem 6.6. *(a) [90] For $d \geq 5$, there is a constant a' depending on d, such that the critical two-point function of the nearest-neighbour strictly self-avoiding walk obeys, as $|x| \to \infty$,*

$$G_{z_c}(x) = \frac{a'}{|x|^{d-2}} \left[1 + O\left(\frac{1}{|x|^{2/d}}\right)\right]. \tag{6.13}$$

(b) [91] Fix any $\alpha > 0$ (think of α small). For $d > 4$, there is a finite constant a depending on d and L, and an L_0 depending on d and α, such that for $L \geq L_0$ the critical two-point function of the spread-out model of strictly self-avoiding walk obeys, as $|x| \to \infty$,

$$G_{z_c}(x) = \frac{a}{|x|^{d-2}} \left[1 + O\left(\frac{1}{|x|^{[2\wedge 2(d-4)]-\alpha}}\right)\right]. \tag{6.14}$$

The constant in the error term is uniform in x but may depend on L and α; a more careful statement in this regard is given in [91].

The asymptotic formula $\xi(z) \sim \sqrt{v/2d}(1 - z/z_c)^{-1/2}$ for the correlation length (2.25) is proved in [97, 98] for the nearest-neighbour model in dimensions $d \geq 5$, using the method of [89]. The same result is obtained in [158] for the spread-out model in dimensions $d > 4$ with L sufficiently large. This confirms the prediction (2.26).

Finally, we mention a result concerning the *infinite* self-avoiding walk, a concept first defined in [147]. Given $n \geq m$ and an m-step self-avoiding walk ω, let $P_{m,n}(\omega)$ denote the fraction of n-step walks which *extend* ω, meaning that the first m steps agree with ω. Then we define

$$P_m(\omega) = \lim_{n \to \infty} P_{m,n}(\omega) \qquad (6.15)$$

if the limit exists. If the limit does exist, then the probability measures P_m on m-step walks will be consistent in the sense of Exercise 2.1. This consistency property allows for the definition via cylinder sets of a measure P_∞ on the set of all infinite self-avoiding walks. The measure P_∞ is the *infinite self-avoiding walk*. In [148], the lace expansion was used to construct the nearest-neighbour infinite self-avoiding walk in dimensions $d \geq d_0$, for some sufficiently large d_0. This was improved in [97] (see also [158]), where the nearest-neighbour infinite self-avoiding walk was constructed in all dimensions $d \geq 5$.

6.2 A Self-Avoiding Walk in Dimension $d = 1$

The above results all consider $d > 4$. An application to 1-dimensional nearest-neighbour weakly self-avoiding walk has been given by van der Hofstad [108], and we discuss this next.

For $\lambda = 1$, the 1-dimensional self-avoiding walk is easy: there are just two self-avoiding walks that start at the origin, one travelling left and one travelling right. However, for $\lambda \in (0,1)$, the problem is much more difficult. It was proved in [84] that the behaviour remains ballistic for all $\lambda \in (0,1)$, i.e., the walk is asymptotically at position θn after n steps, where the speed θ depends on λ. It remains an open problem to prove that θ is an increasing function of λ, although this seems obvious intuitively. The substantial recent progress in the study of 1-dimensional problems of this sort is reviewed in [116].

When λ is close to 1, the weak avoidance is actually rather strong. In the following theorem, due to [108], the lace expansion was used to treat the weakly self-avoiding walk, with λ close to 1, as a small perturbation of the one-dimensional strictly self-avoiding walk. This is the only time that the lace expansion has been used to perturb around a non-Gaussian model.

We replace λ by

$$\lambda = \lambda_{st} = 1 - \exp[-\beta|s - t|^{-p}], \qquad (6.16)$$

with $p \in [0,1]$. For factor $|s - t|^{-p}$ has the effect of diminishing β when $p \in (0,1]$. This corresponds to a "forgetful" self-avoidance, in the sense that the penalty for a self intersection decreases when the time until the self-intersection increases. Let

$$\hat{c}_n^+(k) = \sum_{x \geq 0} c_n(x) e^{ikx}. \tag{6.17}$$

This breaks the left-right symmetry, which is necessary if we are to observe ballistic behaviour.

Theorem 6.7. *Let $d = 1$ and $p \in [0,1]$, and consider the nearest-neighbour model with $\lambda = \lambda_{st}$ given by (6.16). There is a (large) β_0 such that for all $\beta \geq \beta_0$ there exists $v = v(\beta, p)$ and $\theta = \theta(\beta, p)$ such that*

$$\lim_{n \to \infty} e^{-i\theta nk/v\sqrt{n}} \; \frac{\hat{c}_n^+(k/v\sqrt{n})}{\hat{c}_n^+(0)} = e^{-k^2/2}. \tag{6.18}$$

The first factor on the left hand side of (6.18) is an indication of ballistic behaviour with speed θ. Extensions of (6.18) are given in [108]. See [138] for earlier work on this model. It was shown in [113] that if $p > \frac{3}{2}$ and λ is small then the ballistic behaviour of Theorem 6.7 is replaced by diffusive behaviour. Arguments were put forward in [47] that for $\lambda > 0$ and $1 \leq p \leq \frac{3}{2}$ the mean-square displacement is asymptotic to a multiple of $n^{2\nu_p}$, where ν_p varies linearly in p between $\nu_1 = 1$ and $\nu_{3/2} = \frac{1}{2}$. Extensions to higher dimensions are also discussed in [47].

6.3 Nearest-Neighbour Attraction

The self-avoiding walk models a single polymer in a good solution. In a bad solution, it is energetically more favourable for the polymer to be in contact with itself rather than with the solution. This is modelled as follows.

We fix $\kappa > 0$ and replace U_{st} of (2.1) by

$$V_{st}(\omega) = \begin{cases} -\lambda & \text{if } \omega(s) = \omega(t), \\ \kappa & \text{if } \|\omega(s) - \omega(t)\|_1 = 1, \\ 0 & \text{otherwise.} \end{cases} \tag{6.19}$$

Define $D : \mathbb{Z}^d \to \mathbb{R}$ by $D(0) = 0$, $D(x) = a_L e^{-|x|/L}$ for $x \neq 0$, where a_L is chosen so that $\sum_{x \in \mathbb{Z}^d} D(x) = 1$. (In fact, the theorem below, due to [204], allows for a more general choice of D.) For an n-step walk ω, let $W(\omega) = \prod_{j=1}^{n} D(\omega(j) - \omega(j-1))$. Let

$$c_n^{(\lambda,\kappa)}(x) = \sum_{\omega \in \mathcal{W}_n(0,x)} W(\omega) \prod_{0 \leq s < t \leq n} (1 + V_{st}(\omega)), \tag{6.20}$$

where the sum is over n-step walks taking *arbitrary* steps. Let $c_n^{(\lambda,\kappa)} = \sum_{x \in \mathbb{Z}^d} c_n^{(\lambda,\kappa)}(x)$. Then we define a measure by

$$\mathbb{E}_n^{(\lambda,\kappa)} X = \frac{1}{c_n^{(\lambda,\kappa)}} \sum_{\omega \in \mathcal{W}_n} X(\omega) W(\omega) \prod_{0 \leq s < t \leq n} (1 + V_{st}(\omega)). \qquad (6.21)$$

In (6.19), there is a competition between two tendencies. The self-avoidance tends to make the walk more spread out, whereas the nearest-neighbour attraction tends to make the walk more compact. The general picture [132, 205] is that as κ is increased, with λ fixed, there is a collapse transition at a critical value $\kappa_c = \kappa_c(\lambda)$. In this transition the walk's length scale (measured e.g. by average end-to-end distance) jumps from being of order n^ν with ν given by (2.16) when $\kappa < \kappa_c$, to order $n^{1/d}$ for $\kappa > \kappa_c$. The critical value κ_c is known as the theta point, and at the theta point it is predicted that the two competing effects cancel each other, so that (for $d \geq 3$) the length scale is the Gaussian length scale $n^{1/2}$. For simulations of the collapse transition in dimension $d = 5$, see [174].

The following theorem of Ueltshi [204] proves that there is an uncollapsed phase in the above model, for $d > 4$, $\lambda = 1$ and small κ. This theorem is noteworthy in its application to a model for which the interaction is not purely repulsive. The attractive nature of the interaction is problematic, as the inequality $1 + \mathcal{U}_{st} \leq 1$ used in the proof of existence of the connective constant, and of Proposition 4.2, can now be violated. Specific properties of the choice of D above are used in the proof of the theorem, to surmount this difficulty.

Theorem 6.8. *Fix[2] $d > 4$, $\delta \in (0, 1 \wedge \frac{d-4}{2})$, and let V be as above. There is an L_0 and a κ_0 such that for all $L \geq L_0$ and $0 < \kappa \leq \kappa_0$ there are positive constants $A = A(\kappa, L, d)$, $\mu = \mu(\kappa, L, d)$ and $v = v(\kappa, L, d)$ such that*

$$\hat{c}_n^{(1,\kappa)}(0) = A\mu^n [1 + O(n^{-(d-4)/2})], \qquad (6.22)$$
$$\mathbb{E}_n^{(1,\kappa)} |\omega(n)|^2 = vn[1 + O(n^{-\delta})]. \qquad (6.23)$$

The proof makes use of the general inductive theorem of [120], and, as such, has consequences that go beyond (6.22)–(6.23).

6.4 Networks of Self-Avoiding Walks

Polymer networks can be modelled by networks of mutually-avoiding self-avoiding walks. The scaling behaviour of such networks has been studied in the physics literature for general dimensions [66, 67], but mathematically rigorous results are restricted to $d > 4$. Networks in dimensions $d > 4$ having the topology of a tree were studied in [124]. A special case of the results of [124] pertains to star polymers, which we discuss now.

Let $r \geq 1$, $\vec{n} = (n_1, \ldots, n_r)$ with each n_j a positive integer, and $\vec{x} = (x_1, \ldots, x_r)$ with each $x_j \in \mathbb{Z}^d$. Let $s_{\vec{n}}^{(r)}(\vec{x})$ denote the number of *star polymers* consisting of r self-avoiding walks of length $n_1, \ldots, n_r$ starting at the

---

[2] The restriction on δ in [120, Theorem 1.1] contains a misprint and should involve $\frac{d-4}{2}$ rather than $\frac{d-4}{4}$.

Fig. 6.1. Star polymers with $r = 2, 3, 4$.

origin and ending at $x_1, \ldots, x_r$ respectively, which are also *mutually* avoiding apart from their common beginning at the origin. We consider the model in which the walks take steps in the spread-out set Ω of (1.2). See Fig. 6.1. Let $s_{\vec{n}}^{(r)} = \sum_{\vec{x} \in \mathbb{Z}^{dr}} s_{\vec{n}}^{(r)}(\vec{x})$.

Theorem 6.9. *Let[3] $d > 4$, $\delta \in (0, 1 \wedge \frac{d-4}{2})$, $r \geq 1$, $\vec{n} = (n_1, \ldots, n_r)$, and $n = \sum_{j=1}^{r} n_j$, and assume that $n_j \sim n t_j$ with $t_j \in (0, 1]$ for each j. There is an $L_0 = L_0(d)$ such that for $L \geq L_0$ there exist positive constants v, μ, A, V_r (all depending on d and L) with $V_1 = 1$ and $V_2 = A^{-1}$, and there exist C_1, C_2 (depending on d but not L), such that the following statements hold as $n \to \infty$:*

$$s_{\vec{n}}^{(r)} = V_r A^r \mu^n [1 + O(n^{-(d-4)/2})], \tag{6.24}$$

$$\frac{1}{s_{\vec{n}}^{(r)}} \sum_{\vec{x} \in \mathbb{Z}^{dr}} |x_j|^2 s_{\vec{n}}^{(r)}(\vec{x}) = v n_j [1 + O(n^{-\delta})] \quad (j = 1, \ldots, r), \tag{6.25}$$

$$C_1 \mu^n L^{-dr} n^{-dr/2} \leq \sup_{\vec{x} \in \mathbb{Z}^{dr}} s_{\vec{n}}^{(r)}(\vec{x}) \leq C_2 \mu^n L^{-dr} n^{-dr/2}. \tag{6.26}$$

Since (6.24) with $r = 1$ gives $s_n^{(1)} = c_n = A\mu^n [1 + O(n^{-(d-4)/2})]$, μ must be the connective constant given by $\mu = \lim_{n \to \infty} c_n^{1/n}$. Thus the connective constant μ also serves as the growth constant for star polymers. For $d = 3$, a closely related result is proved in [194]; the proof extends to general $d \geq 2$ [193]. Also, (6.25) gives $c_n^{-1} \sum_{x \in \mathbb{Z}^d} |x|^2 c_n(x) = vn[1 + O(n^{-\delta})]$, so v is the diffusion constant.

Theorem 6.9 states that the constants A and V_2 are related by $V_2 = A^{-1}$. In fact, this is required for consistency of (6.24). To see this, consider the statement of (6.24) for $r = 1$ and $n = 2m$. In this case, (6.24) gives $c_{2m} \sim A\mu^{2m}$. On the other hand, we may regard a single polymer of length $2m$ as a star polymer consisting of two branches of length m. With this interpretation, (6.24) gives $c_{2m} \sim A^2 V_2 \mu^{2m}$. Therefore it must be the case that $V_2 = A^{-1}$.

The constants V_r are referred to as *vertex factors*. It is proved in [124] that $V_r = 1 + O(L^{-d})$ for $r \geq 2$. Although not explicitly stated in [124], it follows easily from the results of [124] that there is a positive constant c_r such that

$$V_r \leq 1 - c_r L^{-d} \quad (r \geq 2). \tag{6.27}$$

---

[3] The footnote to Theorem 6.8 applies also here.

The vertex factors take into account the local effect of the mutual avoidance of the walks that meet at the origin, and (6.27) reflects the natural fact that the mutual avoidance diminishes the number of allowed configurations. This is analogous to (5.76), and, in fact, it is the case that $V(z_c)$ of (5.76) is equal to V_2.

Since $s_n^{(1)}(x) = c_n(x)$, (6.26) provides both upper and lower bounds for $c_n(x)$ of the form $n^{-d/2}$. This is a kind of local central limit theorem (cf. Theorems 6.3 and 6.5).

In [126], general graphical networks were considered, possibly containing closed loops. The results of [126] are stated in terms of critical generating functions. For simplicity, we state the result of [126] as it applies to the example of *watermelon networks*. See Fig. 6.2. For $r \geq 2$, let

$$W_z^{(r)}(x) = \sum_{(\omega_1,\ldots,\omega_r)\in\mathcal{S}^{(r)}(0,x)} z^{|\omega_1|+\cdots+|\omega_r|}, \tag{6.28}$$

where the sum is over all r-tuples of self-avoiding walks from 0 to x, of any length, which are mutually avoiding apart from their common endpoints 0 and x. The walks are spread-out walks with steps given by (1.2).

Theorem 6.10. *Let $d > 4$ and $r \geq 2$. Fix $\epsilon_1 < (d-4)\wedge 1$. Let z_c be the critical value, let V_r be the vertex factors of Theorem 6.9, and let a be the constant of (6.14). There exists an $L_0 = L_0(d,r)$ such that for $L \geq L_0$, as $|x| \to \infty$,*

$$W_{z_c}^{(r)}(x) = V_r^2 \frac{a^r}{|x|^{r(d-2)}} \left[1 + O\left(\frac{1}{|x|^{\epsilon_1}}\right)\right]. \tag{6.29}$$

Constants in the error term depend on L, d, r and ϵ_1.

Comparing with the asymptotic formula for the critical two-point function in Theorem 6.6, we can interpret Theorem 6.10 as stating that the leading asymptotic behaviour of the watermelon network is that of r independent self-avoiding walks joining 0 to x, apart from the vertex factors associated with the two vertices of the watermelon network. The results of [126] provide a similar conclusion for networks with the topology of an arbitrary graph.

The proofs of Theorems 6.9–6.10 are based on an extension of the notion of laces on an interval, as defined in Sect. 3.3, to laces on a tree. A general theory of such laces is developed in [124].

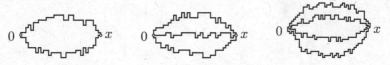

Fig. 6.2. Watermelon networks with $r = 2, 3, 4$.

Lattice Trees

Lattice trees are natural combinatorial objects that fit within the general framework of critical phenomena. They also model branched polymers. In this chapter, we first give an overview of the predicted scaling properties of lattice trees and state results obtained in dimensions $d > 8$ using the lace expansion. We then show that the square condition plays a similar role for lattice trees as the bubble condition does for the self-avoiding walk.

7.1 Asymptotic Behaviour

A *lattice tree* on $\mathbb{Z}^d$ is defined to be a finite connected set of bonds which contains no cycles (closed loops). Bonds are pairs $\{x, y\}$ of vertices of $\mathbb{Z}^d$, with $y - x \in \Omega$, where Ω is given either by the nearest-neighbour set (1.1) or the spread-out set (1.2). Although a tree T is defined as a set of bonds, we will write $x \in T$ if x is an element of a bond in T. The number of bonds in T is denoted $|T|$, and the number of vertices in T is thus $|T| + 1$.

A basic combinatorial problem is to count the number of lattice trees of fixed size. Let $t_n^{(2)}(x)$ denote the number of n-bond lattice trees that contain the two vertices 0 and x, and let $t_n^{(1)} = t_n^{(2)}(0)$ denote the number of n-bond trees that contain the origin. Note that

$$
\hat{t}_n^{(2)}(0) = \sum_{x \in \mathbb{Z}^d} t_n^{(2)}(x) = \sum_{x \in \mathbb{Z}^d} \sum_{T : |T| = n, T \ni 0, x} 1
$$

$$
= \sum_{T : |T| = n, T \ni 0} \sum_{x \in T} 1 = (n+1) t_n^{(1)}(0). \tag{7.1}
$$

It is customary to count lattice trees modulo translation, namely to consider t_n defined by

$$
t_n = \frac{1}{n+1} t_n^{(1)} = \frac{1}{(n+1)^2} \hat{t}_n^{(2)}(0). \tag{7.2}
$$

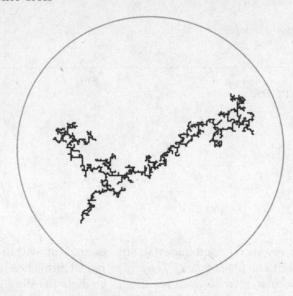

Fig. 7.1. A 2-dimensional lattice tree with $n = 1000$ bonds, generated using the algorithm of [133]. The circle has radius $n^{0.64}$.

A subadditivity argument [145] shows that there is a positive constant λ such that $\lim_{n\to\infty} t_n^{1/n} = \lambda$. The precise asymptotic behaviour of t_n as $n \to \infty$ is believed to be given by

$$t_n \sim \text{const}\lambda^n n^{-\theta}, \tag{7.3}$$

where θ is a universal critical exponent. The bounds

$$c_1 n^{-c_2 \log n}\lambda^n \le t_n \le c_3 n^{-(d-1)/d}\lambda^n, \tag{7.4}$$

proved respectively in [131] and [157], are the best general bounds known at present for t_n.

In studying t_n, it is convenient to introduce the *two-point function*

$$G_z(x) = \sum_{n=0}^{\infty} t_n^{(2)}(x)z^n = \sum_{T:T\ni 0,x} z^{|T|}, \tag{7.5}$$

and the *susceptibility* $\chi(z) = \sum_x G_z(x) = \hat{G}_z(0)$. The susceptibility has radius of convergence $z_c = \lambda^{-1}$. It is a consequence of (7.2)–(7.3) that $\hat{t}_n^{(2)}(0) \sim \text{const}\lambda^n n^{2-\theta}$, and this corresponds, at least formally, to a singularity

$$\chi(z) \sim \text{const}\frac{1}{(1 - z/z_c)^\gamma} \tag{7.6}$$

for the susceptibility, with $\gamma = 3 - \theta$. As in (2.27)–(2.28), it is predicted that

$$G_{z_c}(x) \sim \text{const} \frac{1}{|x|^{d-2+\eta}} \tag{7.7}$$

as $|x| \to \infty$, and that

$$\hat{G}_{z_c}(k) \sim \text{const} \frac{1}{|k|^{2-\eta}} \tag{7.8}$$

as $k \to 0$, where η is universal.

The typical length scale of a lattice tree is characterized by the average radius of gyration R_n, defined as follows. The squared average radius of gyration is defined by

$$R_n^2 = \frac{1}{t_n^{(1)}} \sum_{T:|T|=n, T \ni 0} R(T)^2, \tag{7.9}$$

where

$$R(T)^2 = \frac{1}{|T|+1} \sum_{x \in T} |x - \bar{x}_T|^2 \tag{7.10}$$

is the squared radius of gyration of T. In (7.10), $\bar{x}_T = (|T|+1)^{-1} \sum_{x \in T} x$ denotes the centre of mass of T (considered as a set of equal masses at the *vertices* of T).

Exercise 7.1. Show that

$$R_n^2 = \frac{1}{2\hat{t}_n^{(2)}(0)} \sum_x |x|^2 t_n^{(2)}(x). \tag{7.11}$$

It is predicted that there is a universal critical exponent ν such that

$$R_n \sim \text{const} \, n^\nu. \tag{7.12}$$

It is also believed that the critical exponents γ, η, ν are related by Fisher's relation $\gamma = (2 - \eta)\nu$. In principle, to construct a scaling limit of lattice trees the lattice spacing should be rescaled from 1 to $n^{-\nu}$, followed by the limit $n \to \infty$. It is not immediately obvious how to describe the scaling limit of lattice trees, but a useful description for $d > 8$ will be discussed in Sect. 16.4.

A *lattice animal* is a connected set of bonds which may contain closed loops. It is believed that lattice trees belong to the same universality class as lattice animals, so that lattice trees and lattice animals have the same critical exponents and scaling limits. A definition of weakly self-avoiding lattice trees was given in [30], which presumably is also in the same universality class.

A model of continuous branched polymers, which is also expected to be in the same universality class, was proposed by Brydges and Imbrie and analyzed in [41, 42]. Inspired by ideas of Parisi and Sourlas [175], Brydges and Imbrie [41] proved existence of critical exponents for their continuum model in dimensions $d = 2$ and 3 (with partial results for $d = 4$), with values $\theta = 1$ for

$d = 2$, and $\theta = \frac{3}{2}, \nu = \frac{1}{2}, \eta = -1$ for $d = 3$. It is believed that two-dimensional lattice trees do not have a conformally invariant scaling limit, and that they are not among the substantial class of models described by the Schramm–Loewner evolution SLE_κ. It has been conjectured [36] that η is negative for all dimensions $2 < d < 8$.

Based on a field theoretic representation, it was argued in [155] that the upper critical dimension for lattice trees is 8. Further evidence for this was given in [36,199,201]. The mean-field values of the exponents are $\gamma = \frac{1}{2}, \nu = \frac{1}{4}$ and $\eta = 0$. The value $\nu = \frac{1}{4}$ corresponds in (7.12) to $n \sim \text{const} R_n^4$, which is a statement of 4-dimensionality. The fact that two 4-dimensional objects generically do not intersect in more than eight dimensions gives a quick prediction that $d = 8$ is the upper critical dimension for lattice trees.

The lace expansion has been used to prove a number of results for lattice trees in dimensions $d > 8$. The following theorem [90, 91] shows that $\eta = 0$ for nearest-neighbour lattice trees in sufficiently high dimensions, and for sufficiently spread-out models when $d > 8$. Similar results are proved in [90, 91] for lattice animals.

Theorem 7.2. *(a) [90] There is a d_0 such that for $d \geq d_0$, there is a constant a' depending on d, such that the critical lattice tree two-point function obeys, as $|x| \to \infty$,*

$$G_{z_c}(x) = \frac{a'}{|x|^{d-2}} \left[1 + O\left(\frac{1}{|x|^{2/d}} \right) \right]. \tag{7.13}$$

(b) [91] Let $d > 8$, fix any $\alpha > 0$ (think of α small), and consider the spread-out model of lattice trees. There is a finite constant a depending on d and L, and an L_0 depending on d and α, such that for $L \geq L_0$, as $|x| \to \infty$,

$$G_{z_c}(x) = \frac{a}{|x|^{d-2}} \left[1 + O\left(\frac{1}{|x|^{[2 \wedge (d-8)] - \alpha}} \right) \right]. \tag{7.14}$$

The constant in the error term is uniform in x but may depend on L and α; a more careful statement in this regard is given in [91].

The following theorem, proved in [99], proves that $\theta = \frac{5}{2}$ and $\nu = \frac{1}{4}$ in high dimensions. Weaker results have been obtained for lattice animals, at the level of generating functions [95].

Theorem 7.3. *For nearest-neighbour lattice trees with d sufficiently large, or for spread-out lattice trees with $d > 8$ and L sufficiently large, there are positive constants A and D (depending on d, L) such that for every $\epsilon < \frac{1}{2} \wedge \frac{d-8}{4}$,*

$$t_n = A\lambda^n n^{-5/2}[1 + O(n^{-\epsilon})], \tag{7.15}$$

$$R_n = Dn^{1/4}[1 + O(n^{-\epsilon})]. \tag{7.16}$$

As in the proof of Theorem 6.2, the proof of Theorem 7.3 proceeds first by studying the singularity of $\hat{G}_z(k)$ at $z = z_c = \lambda^{-1}$, and then uses complex variable methods to extract the asymptotics of t_n and the radius of gyration. In particular, it is shown that under the hypotheses of the theorem the susceptibility obeys

$$\chi(z) = \frac{A\sqrt{\pi}}{\sqrt{1 - z/z_c}} + O(|1 - z/z_c|^{-1/2+\epsilon}) \tag{7.17}$$

for all complex $|z| \le z_c$, which implies that γ exists and equals $\frac{1}{2}$.

The scaling limit of lattice trees in dimensions $d > 8$ will be discussed in Sect. 16.4, but the following theorem of [62] gives a first step. For the statement, let A and D be the constants of Theorem 7.3, and let

$$D_0 = 2^{3/4}\pi^{-1/4}D. \tag{7.18}$$

Theorem 7.4. *For nearest-neighbour lattice trees with d sufficiently large, or for spread-out lattice trees with $d > 8$ and L sufficiently large, as $n \to \infty$,*

$$\hat{t}_n^{(2)}(kD_0^{-1}n^{-1/4}) \sim A\lambda^n n^{-1/2} \int_0^\infty dt\, t\, e^{-t^2/2}e^{-|k|^2t/2d}. \tag{7.19}$$

The integral on the right hand side of (7.19) will be discussed in Chap. 16. Spatial scaling by $n^{-1/4}$ corresponds to replacement of the Fourier variable k by $kn^{-1/4}$, and (7.19) is a statement about the scaling limit of $t_n^{(2)}(x)$ in Fourier language. Note that (7.15) can be combined with (7.2) to give $\hat{t}_n^{(2)}(0) \sim A\lambda^n n^{-1/2}$, which is consistent with (7.19).

We will not prove Theorems 7.2–7.4 here. The proofs involve deriving a lace expansion for lattice trees, and diagrammatic estimates for the expansion, followed by proof of convergence of the expansion. The latter can be handled in various ways [91, 95, 99, 125]. Significant new issues do arise in the proofs of Theorems 7.3–7.4. The modifications needed to apply the lace expansion to lattice trees will be discussed in Chap. 8. But first, we indicate how the role of the bubble condition for self-avoiding walks is replaced by the square condition for lattice trees.

7.2 Differential Inequalities and the Square Condition

In this section, we adapt the method and results of Sect. 2.2 to lattice trees.

For $r \ge 1$, we define the r-point function by

$$G_z^{(r)}(x_1, \ldots, x_{r-1}) = \sum_{T:T \ni 0, x_1, \ldots, x_{r-1}} z^{|T|}. \tag{7.20}$$

In particular, $G_z^{(2)}(x)$ is the two-point function $G_z(x)$ of (7.5), and we will write $g_z = \sum_{T:T\ni 0} z^{|T|}$ for the one-point function. For $r \geq 1$,

$$\frac{\mathrm{d}[zG_z^{(r)}(x_1,\ldots,x_{r-1})]}{\mathrm{d}z} = \sum_{T:T\ni 0,x_1,\ldots,x_{r-1}} (|T|+1)z^{|T|}$$

$$= \sum_{x_r} G_z^{(r+1)}(x_1,\ldots,x_r). \tag{7.21}$$

For $r = 1, 2$, this implies that

$$\frac{\mathrm{d}[zg_z]}{\mathrm{d}z} = \sum_{x\in\mathbb{Z}^d} G_z(x) = \chi(z), \tag{7.22}$$

and

$$\frac{\mathrm{d}[z\chi(z)]}{\mathrm{d}z} = \sum_{x,y} G_z^{(3)}(x,y). \tag{7.23}$$

We seek upper and lower bounds for the right hand side of (7.23), which will provide differential inequalities for the susceptibility.

For an upper bound, we note that a lattice tree containing vertices $0, x, y$ contains a unique *skeleton* consisting of the union of the paths in the tree joining 0 to x, and 0 to y. The skeleton divides naturally into three paths, which join each of $0, x, y$ to the vertex u (say) where the path from 0 to y separates from the path from 0 to x. By neglecting the mutual avoidance of the three subtrees defined naturally by these paths (the "inclusion" part of inclusion-exclusion), we obtain

$$G_z^{(3)}(x,y) \leq \sum_u G_z(u)G_z(x-u)G_z(y-u). \tag{7.24}$$

Inserting this into (7.23) then gives the upper bound

$$\frac{\mathrm{d}[z\chi(z)]}{\mathrm{d}z} \leq \chi(z)^3. \tag{7.25}$$

For a lower bound, we first define the *square diagram* by

$$\mathsf{S}(z) = \sum_{x,y,u} G_z(x)G_z(y-x)G_z(u-y)G_z(u). \tag{7.26}$$

Using the inverse Fourier transform, it follows from Exercise 1.1 that

$$\mathsf{S}(z) = \int_{[-\pi,\pi]^d} \hat{G}_z(k)^4 \frac{\mathrm{d}^d k}{(2\pi)^d}. \tag{7.27}$$

The *square condition* states that $\mathsf{S}(z_c) < \infty$, and will be satisfied for $d > 8$ if $\eta = 0$ in (7.7)–(7.8). Keeping only the term $x = y = u = 0$ in (7.26), we see that $\mathsf{S}(z) \geq g_z^4 \geq g_z$, and hence the square condition implies that $g_{z_c} < \infty$.

For the "exclusion" part of the inclusion-exclusion, we note two sources of overcounting in (7.24). One source is the fact that each two-point function on the right hand side of (7.24) contributes a branch at u, and this overcounting can be handled by inserting a factor $1/g_z^2$ to prune two of the branches. Another source arises from the possible intersection of the three subtrees. Bounding this subtracted term from above, we are led to the lower bound depicted in Fig. 7.2 (see [95, Sect. 1.3] for more details). Summing this lower bound gives

$$\chi(z)^3 \left[\frac{1}{g_z^2} - 3[S(z) - g_z^4] \right] \leq \frac{\mathrm{d}[z\chi(z)]}{\mathrm{d}z} \leq \chi(z)^3. \tag{7.28}$$

The diagrams in Fig. 7.2 have a dual interpretation. On the one hand, they give a schematic depiction of actual tree configurations (or of tree-like configurations when self-intersections are imposed on the left hand side). On the other hand, the diagrams have the interpretation that each line represents a two-point function, with unlabelled vertices summed over $\mathbb{Z}^d$. The latter is an upper bound on generating functions of the actual configurations, and it is the latter interpretation that is literally intended in Fig. 7.2. It is nevertheless instructive to keep the former interpretation in mind also.

Exercise 7.5. Fill in the missing details of the derivation of the lower bound of (7.28).

The skeleton inequalities were derived and used to prove the following theorem in [36, 199, 201].

Theorem 7.6. *For all d, $\chi(z) \geq const(1 - z/z_c)^{-1/2}$ for $0 \leq z \leq z_c$. If the square condition is satisfied then the reverse inequality also holds, and hence*

$$\chi(z) \simeq (1 - z/z_c)^{-1/2}. \tag{7.29}$$

Proof. We first discuss the integration of the upper bound of (7.28), to obtain the lower bound in the statement of the theorem. Fix some $z_0 \in (0, z_c)$. There is no difficulty for $z < z_0$. For $z \geq z_0$, (7.28) implies that

$$\frac{\mathrm{d}[z\chi(z)]}{\mathrm{d}z} \leq \frac{1}{z_0^3}[z\chi(z)]^3 \tag{7.30}$$

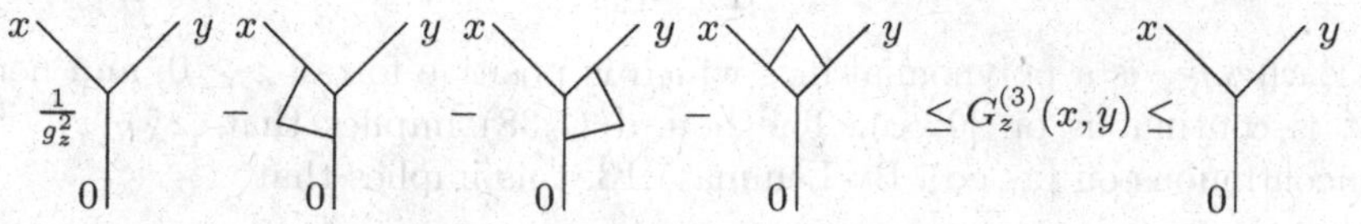

Fig. 7.2. The "skeleton inequalities" (7.28) for lattice trees.

and hence

$$-\frac{\mathrm{d}[z\chi(z)]^{-2}}{\mathrm{d}z} \le \frac{2}{z_0^3}. \tag{7.31}$$

Integration of this inequality over the interval $[z, z_c]$ gives

$$[z\chi(z)]^{-2} - [z_c\chi(z_c)]^{-2} \le \frac{2}{z_0^3}(z_c - z). \tag{7.32}$$

If we knew that $\chi(z_c) = \infty$, this would give the desired lower bound on the susceptibility. Since z_c is the radius of convergence of χ, we do know that $\chi(z) = \infty$ for $z > z_c$, and we need to rule out the possibility of a jump at z_c.

To do so, we argue as follows. Consider lattice trees which lie in the finite set Λ_R given by the intersection of $[-R, R]^d$ with $\mathbb{Z}^d$; the lack of translation invariance of Λ_R complicates the argument. For $x, y \in \Lambda_R$, let

$$G_{R,z}(x, y) = \sum_{T \subset \Lambda_R : T \ni x, y} z^{|T|}, \tag{7.33}$$

$$\chi_{R,x}(z) = \sum_{y \in \Lambda_R} G_{R,z}(x, y), \tag{7.34}$$

$$\bar{\chi}_R(z) = \max_{x \in \Lambda_R} \chi_{R,x}(z). \tag{7.35}$$

Note that

$$
\begin{aligned}
\chi(z) &= \lim_{R \to \infty} \chi_{R,0}(z) = \sup_R \chi_{R,0}(z) \\
&= \lim_{R \to \infty} \bar{\chi}_R(z) = \sup_R \bar{\chi}_R(z).
\end{aligned} \tag{7.36}
$$

It is not difficult to follow the steps leading to the upper bound of (7.28) to see that for each $x \in \Lambda_R$,

$$\frac{\mathrm{d}[z\chi_{R,x}(z)]}{\mathrm{d}z} \le \chi_{R,x}(z)\bar{\chi}_R(z)^2 \le \bar{\chi}_R(z)^3. \tag{7.37}$$

Since each $\chi_{R,x}$ is a polynomial, and since there are finitely many $x \in \Lambda_R$, for each z there is an x such that the derivative of $\chi_{R,x}$ is equal to the derivative of $\bar{\chi}_R$, except possibly for finitely many values of z. Apart from this exceptional set, it follows from (7.37) that

$$-\frac{\mathrm{d}[z\bar{\chi}_R(z)]^{-2}}{\mathrm{d}z} \le \frac{2}{z^3}. \tag{7.38}$$

Each $\chi_{R,x}$ is a polynomial in z which is positive for all $z \ge 0$, and hence $\bar{\chi}_R^{-2}$ is continuous on $[0, \infty)$. The bound (7.38) implies that $[z\bar{\chi}_R(z)]^{-2}$ is equicontinuous on $[z_0, \infty)$. By Lemma 5.13, this implies that

$$[z\chi(z)]^{-2} = \inf_R [z\bar{\chi}_R(z)]^{-2} = -\sup_R \left(-[z\bar{\chi}_R(z)]^{-2} \right) \tag{7.39}$$

is continuous on $[z_0, \infty)$. Since $\chi(z)^{-2} = 0$ for $z > z_c$, it follows that $\chi(z_c)^{-2} = 0$, as required.

For the upper bound, we will only observe here the weaker result that if $S(z_c) - g_{z_c}^4 < (3g_{z_c}^2)^{-1}$, then the lower bound of (7.30) can be integrated to give the upper bound on χ stated in the theorem. Since $g_{z_c}^2 \leq S(z_c)$ by definition, the above inequality is implied if $S(z_c)(S(z_c) - g_{z_c}^4) < \frac{1}{3}$, which follows if $S(z_c) - g_{z_c}^4$ is sufficiently small. In applications of the lace expansion, it is always proved that $S(z_c) - g_{z_c}^4$ is small, so the weaker result suffices in practice. However, the method used in [12], to deal with the analogous issue for percolation, can actually be applied to show that in fact $S(z_c) < \infty$ is sufficient to imply the desired upper bound on the susceptibility [199].   ∎

By analogy with the discussion of Sect. 5.4, it might be expected that for $d > 8$ there is a constant v such that

$$\frac{\mathrm{d}[z\chi(z)]}{\mathrm{d}z} \sim v\chi(z)^3 \qquad \text{as } z \to z_c^-. \tag{7.40}$$

In fact, the lace expansion has been used to prove that under the hypotheses of Theorem 7.3,

$$\frac{\mathrm{d}[z\chi(z)]}{\mathrm{d}z} = \frac{1}{2\pi A^2}\chi(z)^3 + O(|1 - z/z_c|^{-3/2+\epsilon})$$

$$= \frac{A\sqrt{\pi}}{2(1 - z/z_c)^{3/2}} + O(|1 - z/z_c|^{-3/2+\epsilon}) \tag{7.41}$$

holds uniformly in complex z with $|z| < z_c$, where A is the constant of (7.15) and (7.17). The proof of (7.41) in [99] actually uses a *double* expansion, although the lace expansion on a tree derived in [124] can now be used instead to do it with a single expansion [125].

The asymptotic formula (7.41) implies (7.15) via a Tauberian theorem, as the following exercise shows.

Exercise 7.7. (a) Suppose that $f(z) = \sum_{n=0}^{\infty} a_n z^n$ has radius of convergence 1. Suppose that $|f(z)| \leq \text{const} |1 - z|^{-b}$ uniformly in $|z| < 1$, with $b \geq 1$. Prove that $|a_n| \leq \text{const } n^{b-1}$ if $b > 1$, and that $|a_n| \leq \text{const} \log n$ if $b = 1$. Hint:

$$a_n = \frac{1}{2\pi i} \oint_\Gamma f(z) \frac{\mathrm{d}z}{z^{n+1}}$$

where Γ is the contour $|z| = 1 - \frac{1}{n}$. (See [75] or [62] for a solution.)
(b) Conclude (7.15) from (7.41).

8

The Lace Expansion for Lattice Trees

In this chapter, we derive the lace expansion for lattice trees. We also indicate how the expansion can be modified to apply to lattice animals, as this gives some insight into the expansion for percolation to be discussed later in Chap. 10. Finally, we briefly discuss diagrammatic estimates for lattice trees. The ideas in this chapter are from [95]. Convergence of the expansion will not be discussed here; see [91, 95, 125] for three different methods to prove convergence.

8.1 Expansion for Lattice Trees

We first discuss the derivation of the expansion using connected graphs and laces, and then briefly indicate how to interpret the expansion as arising from repeated inclusion-exclusion.

Given two distinct vertices x, y and a tree $T \ni x, y$, the *backbone* is defined to be the unique path, consisting of bonds of T, which joins x to y. If the backbone consists of n bonds, and is considered to start at x and end at y, then removal of the bonds in the backbone disconnects T into $n + 1$ mutually nonintersecting trees $R_0, ..., R_n$, which we refer to as *ribs*. This decomposition is shown in Fig. 8.1.

Given a set $\vec{R} = \{R_0, \ldots, R_n\}$ of $n + 1$ trees R_j, we define

$$U_{st}(\vec{R}) = \begin{cases} -1 & \text{if } R_s \text{ and } R_t \text{ share a common vertex} \\ 0 & \text{if } R_s \text{ and } R_t \text{ share no common vertex.} \end{cases} \quad (8.1)$$

Then the two-point function can be written

$$G_z(x) = \sum_{\omega \in \mathcal{W}(0,x)} z^{|\omega|} \sum_{R_0 \ni 0} \cdots \sum_{R_{|\omega|} \ni x} z^{|R_0| + \ldots + |R_{|\omega|}|} \prod_{0 \leq s < t \leq |\omega|} (1 + U_{st}(\vec{R})),$$

$$(8.2)$$

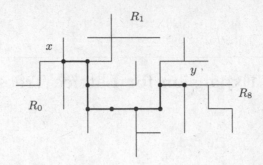

Fig. 8.1. Decomposition of a tree T containing vertices x and y into its backbone and ribs $R_0, \ldots, R_8$. The vertices of the backbone are indicated by heavy dots.

where the sum over ω is over all simple random walks from 0 to x, and where each sum over R_i is a sum over trees containing $\omega(i)$, and $\vec{R} = (R_0, \ldots, R_{|\omega|})$. We will abbreviate the right hand side of (8.2) as

$$G_z(x) = \sum_{\omega \in \mathcal{W}(0,x)} z^{|\omega|} \left[\prod_{i=0}^{|\omega|} \sum_{R_i \ni \omega(i)} z^{|R_i|} \right] \prod_{0 \le s < t \le |\omega|} (1 + U_{st}(\vec{R})), \qquad (8.3)$$

with the understanding that the sums over R_i cannot be completed in (8.3) without taking into account the product over s, t. The identity (8.2) represents the generating function for lattice trees containing 0 and x by a sum which generates backbones ω and ribs $\vec{R}$, with an interaction to prevent the ribs from intersecting each other. The sum over ω is the sum over all random walks from 0 to x (taking steps in Ω), although walks that are not self-avoiding give zero contribution to (8.2).

8.1.1 The Expansion

We use the terminology of Definitions 3.2 and 3.5, but with a change in the definition of connected graph. A graph Γ on $[a, b]$ is now be said to be *connected* if both a and b are endpoints of edges in Γ, and if in addition for each $c \in (a, b)$ there are $s, t \in [a, b]$ such that $s < c < t$ with either (i) $\{s, t\} \in \Gamma$, or (ii) $\{c, t\} \in \Gamma$ and $\{s, c\} \in \Gamma$. Equivalently, Γ is connected if $\cup_{st \in \Gamma}[s, t] = [a, b]$. This notion of connectedness is less restrictive than that used for the self-avoiding walk, and is better suited for dealing with the interaction between ribs. This new definition of connected graph leads to a larger set of laces than before, where we still define a *lace* to be a minimally connected graph, i.e., a connected graph for which the removal of any edge would result in a disconnected graph. The set of laces on $[a, b]$ is denoted $\mathcal{L}[a, b]$. We also modify the prescription associating to each connected graph Γ a unique lace L_Γ, to conform with the new notion of connectedness. Namely, we now define L_Γ to consist of edges $s_1 t_1, s_2 t_2, \ldots$, with $t_1, s_1, t_2, s_2, \ldots$ determined, in that order, by

$$t_1 = \max\{t : at \in \Gamma\}, \quad s_1 = a,$$

$$t_{i+1} = \max\{t : \exists s \leq t_i \text{ such that } st \in \Gamma\}, \quad s_{i+1} = \min\{s : st_{i+1} \in \Gamma\}.$$

As in Sect. 3.3, given a lace L, we define $\mathcal{C}(L) = \{st \notin L : \mathsf{L}_{L \cup \{st\}} = L\}$.
Let

$$K[a, b] = \prod_{a \leq s < t \leq b} (1 + U_{st}), \tag{8.4}$$

and, as in (3.9) and (3.19),

$$
\begin{aligned}
J[0, a] &= \sum_{\Gamma \in \mathcal{G}[a,b]} \prod_{st \in \Gamma} U_{st} \\
&= \sum_{L \in \mathcal{L}[0,a]} \prod_{st \in L} U_{st} \prod_{s't' \in \mathcal{C}(L)} (1 + U_{s't'}),
\end{aligned} \tag{8.5}
$$

where now $\mathcal{G}[a, b]$ is the set of connected graphs with the new definition of connectivity.

Exercise 8.1. Modify the proof of Lemma 3.4 to show that with the new definition of connectivity,

$$K[0, b] = K[1, b] + \sum_{a=1}^{b-1} J[0, a] K[a + 1, b] + J[0, b], \tag{8.6}$$

for $b \geq 1$ (the empty sum is 0 if $b = 1$).

Substitution of (8.6) into (8.3) results in

$$
\begin{aligned}
G_z(x) &= \sum_{R_0 \ni 0} z^{|R_0|} \delta_{0,x} + \sum_{\substack{\omega \in \mathcal{W}(0, x) \\ |\omega| \geq 1}} z^{|\omega|} \left[\prod_{i=0}^{|\omega|} \sum_{R_i \ni \omega(i)} z^{|R_i|} \right] K[1, |\omega|] \\
&\quad + \sum_{\substack{\omega \in \mathcal{W}(0, x) \\ |\omega| \geq 2}} z^{|\omega|} \left[\prod_{i=0}^{|\omega|} \sum_{R_i \ni \omega(i)} z^{|R_i|} \right] \sum_{a=1}^{|\omega|-1} J[0, a] K[a + 1, |\omega|] \\
&\quad + \sum_{\substack{\omega \in \mathcal{W}(0, x) \\ |\omega| \geq 1}} z^{|\omega|} \left[\prod_{i=0}^{|\omega|} \sum_{R_i \ni \omega(i)} z^{|R_i|} \right] J[0, |\omega|]. \tag{8.7}
\end{aligned}
$$

The first term on the right hand side is due to the contribution to (8.3) from the zero-step walk, and the other terms are due to the walks ω with $|\omega| \geq 1$.

Recalling our notation g_z for the one-point function, and writing

$$\Pi_z(x) = \sum_{\substack{\omega \in \mathcal{W}(0,x) \\ |\omega| \geq 1}} z^{|\omega|} \left[\prod_{i=0}^{|\omega|} \sum_{R_i \ni \omega(i)} z^{|R_i|} \right] J[0,|\omega|], \qquad (8.8)$$

the first and last terms on the right hand side of (8.7) are equal to $g_z \delta_{0,x}$ and $\Pi_z(x)$, respectively. The second term on the right hand side of (8.7) is equal to

$$\sum_{R_0 \ni 0} z^{|R_0|} z \sum_{u \in \Omega} G_z(x - u) = g_z(z|\Omega|D * G_z)(x), \qquad (8.9)$$

where D is defined by (1.10). For the third term on the right hand side of (8.7), we consider ω to be composed of an initial walk ω_1 from 0 to (say) u, followed by a single step to (say) v, and then a final portion ω_2 from v to x. The term in question is then equal to

$$z \sum_{(u,v)} \sum_{\substack{\omega_1 \in \mathcal{W}(0,u) \\ |\omega_1| \geq 1}} z^{|\omega_1|} \left[\prod_{i=0}^{|\omega_1|} \sum_{R_i \ni \omega_1(i)} z^{|R_i|} \right] J[0,|\omega_1|]$$

$$\times \sum_{\substack{\omega_2 \in \mathcal{W}(v,x) \\ |\omega_2| \geq 0}} z^{|\omega_2|} \left[\prod_{i=0}^{|\omega_2|} \sum_{R_i \ni \omega_2(i)} z^{|R_i|} \right] K[0,|\omega_2|]$$

$$= z \sum_{(u,v)} \Pi_z(u) G_z(x - v) = (\Pi_z * z|\Omega|D * G_z)(x), \qquad (8.10)$$

where the sum over (u,v) denotes the sum over all directed steps with $v - u \in \Omega$.

Summarizing, (8.7) can be rewritten as

$$G_z(x) = g_z \delta_{0,x} + \Pi_z(x) + g_z(z|\Omega|D * G_z)(x) + (\Pi_z * z|\Omega|D * G_z)(x). \quad (8.11)$$

This is the lace expansion for lattice trees. We can write this more compactly by defining

$$h_z(x) = g_z \delta_{0,x} + \Pi_z(x). \qquad (8.12)$$

Then (8.11) becomes

$$G_z(x) = h_z(x) + (h_z * z|\Omega|D * G_z)(x). \qquad (8.13)$$

It is sometimes useful to modify the above analysis by introducing a "fugacity" ζ associated to backbone length. Thus, we define

$$G_{z,\zeta}(x) = \sum_{T:T \ni 0,x} z^{|T|} \zeta^{|\omega_T(x)|}, \qquad (8.14)$$

where $\omega_T(x)$ denotes the backbone joining 0 to x in T. This is the same as inserting a factor $\zeta^{|\omega|}$ in (8.3). Following the steps of the expansion, we obtain

$$G_{z,\zeta}(x) = h_{z,\zeta}(x) + (h_{z,\zeta} * z\zeta|\Omega|D * G_{z,\zeta})(x), \qquad (8.15)$$

where

$$h_{z,\zeta}(x) = g_z \delta_{0,x} + \Pi_{z,\zeta}(x) \qquad (8.16)$$

with $\Pi_{z,\zeta}$ given by (8.8) with an additional factor $\zeta^{|\omega|}$ inserted in the right hand side.

8.1.2 Inclusion–Exclusion

We now interpret the expansion (8.11) as the result of repeated inclusion-exclusion.

The case $|\omega| = 0$ in (8.2) gives rise to the term $g_z \delta_{0,x}$ in (8.11). Consider the case $|\omega| \geq 1$, and suppose that in (8.2) we neglect the fact that the rib R_0 at the origin should avoid the other ribs. This is the same as replacing the interaction $K[0, |\omega|]$ by $K[1, |\omega|]$. The result can be factorized, and produces the term $g_z(z|\Omega|D * G_z)(x)$ on the right hand side of (8.11).

In neglecting the fact that R_0 should avoid the other ribs, we have included terms in which a rib R_j with $j \geq 1$ intersects R_0, and we should exclude their contribution by subtracting a correction term. Given a configuration in which some R_j with $j \geq 1$ intersects R_0, let j_0 be the smallest value of j for which R_j intersects R_0. As a first approximation, we neglect the fact that the ribs R_j with $j > j_0$ should avoid those with $1 \leq j \leq j_0$. In this approximation, the subtracted term is the convolution with $(z|\Omega|D * G_z)$ of a generating function for tree-like configurations in which the first and last ribs intersect each other. To correct for the approximation, a term involving further rib intersections must be added, and so on. This leads to an identity of the form (8.11), in which Π_z is represented by an alternating series. This alternating series will appear in more detail when diagrammatic estimates are discussed in Sect. 8.3.

8.2 Expansion for Lattice Animals

A *lattice animal* on $\mathbb{Z}^d$ is defined to be a finite connected set of bonds, which may or may not contain cycles. Lattice animals are predicted to have the same critical exponents and scaling limit as lattice trees.

The lace expansion for lattice trees can be modified to apply to lattice animals. The main difference compared to lattice trees is that a lattice animal containing vertices x and y need not contain a *unique* path joining x and y. To deal with this, some definitions are needed.

A lattice animal A containing x and y is said to have a *double connection* from x to y if there are two bond-disjoint self-avoiding walks in A between x and y (the walks may share a common vertex, but no common bond), or if

$x = y$. A bond $\{u, v\}$ in A is called *pivotal* for the connection from x to y if its removal would disconnect the animal into two connected components with x in one connected component and y in the other. There is a natural order to the set of pivotal bonds for the connection from x to y, and each pivotal bond is ordered in a natural way, as follows. The *first* pivotal bond for the connection from x to y (assuming there is at least one) is the pivotal bond for which there is a double connection between one endpoint of the pivotal bond and x. The endpoint for which there is a double connection to x is then the *first* endpoint of the first pivotal bond. To determine the second pivotal bond, the role of x is then played by the second endpoint of the first pivotal bond, and so on.

Let

$$G_z^a(x) = \sum_{A:A\ni 0,x} z^{|A|} \tag{8.17}$$

denote the two-point function for lattice animals, where $|A|$ is the number of bonds in A. Given two vertices x, y and an animal A containing x and y, the *backbone* of A is now defined to be the set of pivotal bonds for the connection from x to y. In general this backbone is not connected. The *ribs* of A are the connected components which remain after the removal of the backbone from A. An example is depicted in Fig. 8.2. The set of all animals having a double connection between x and y is denoted $\mathcal{D}_{x,y}$, and we write

$$g_z^a(y - x) = \sum_{R \in \mathcal{D}_{x,y}} z^{|R|}. \tag{8.18}$$

In particular $\mathcal{D}_{x,x}$ is the set of all animals containing x. Let B be an arbitrary finite ordered set of directed bonds:

$$B = \left((u_1, v_1), \ldots, (u_{|B|}, v_{|B|}) \right).$$

Let $v_0 = 0$ and $u_{|B|+1} = x$. Then

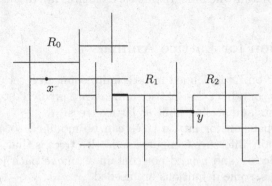

Fig. 8.2. Decomposition of a lattice animal A containing x and y into backbone and ribs. The backbone, consisting of two bonds, is drawn in bold lines.

$$G_z^a(x) = \sum_{B:|B|\geq 0} z^{|B|} \left[\prod_{i=0}^{|B|} \sum_{R_i \in \mathcal{D}_{v_i,u_{i+1}}} z^{|R_i|} \right] K[0,|B|],$$

where $K[0,|B|]$ is defined in (8.4), with U_{st} of (8.1).

Recall the definition of $J[0,a]$ in (8.5), and define

$$\Pi_z^a(y) = \sum_{B:|B|\geq 1} z^{|B|} \left[\prod_{i=0}^{|B|} \sum_{R_i \in \mathcal{D}_{v_i,u_{i+1}}} z^{|R_i|} \right] J[0,|B|] \qquad (8.19)$$

and

$$h_z^a(x) = g_z^a(x) + \Pi_z^a(x). \qquad (8.20)$$

A calculation similar to that used to derive (8.11), using (8.6), gives

$$G_z^a(x) = h_z^a(x) + (h_z^a * z|\Omega|D * G_z^a)(x). \qquad (8.21)$$

This is the lace expansion for lattice animals.

8.3 Diagrammatic Estimates for Lattice Trees

To prove convergence of the lace expansion for lattice trees, bounds are required on $\hat{\Pi}_z(k)$ and $\hat{\Pi}_z(0) - \hat{\Pi}_z(k)$. In this section we sketch the idea of how appropriate diagrammatic bounds can be obtained on $\hat{\Pi}_z(k)$. The procedure for obtaining bounds for lattice animals is similar but more involved, and will not be discussed here. For further details on diagrammatic estimates, both for lattice trees and lattice animals, see [91, 95].

We will not discuss the actual convergence proof here. One approach to convergence, based on Lemma 5.9 and related in spirit to the proof for the self-avoiding walk given in Sect. 5.2, can be found in [95]. A second approach, also based on Lemma 5.9, works directly in x-space and does not use Fourier transforms [91]. A third approach, based on the induction method of [120], is given in [125].

For the diagrammatic estimates for lattice trees, we recall the (new) notions of connectivity and lace discussed in Sect. 8.1.1. We take U_{st} to be defined by (8.1), and we let $\mathcal{L}^{(N)}[a,b]$ denote the set of laces in $\mathcal{L}[a,b]$ consisting of exactly N edges. Let

$$J^{(N)}[a,b] = \sum_{L \in \mathcal{L}^{(N)}[a,b]} \prod_{st \in L} U_{st} \prod_{s't' \in \mathcal{C}(L)} (1 + U_{s't'}) \qquad (8.22)$$

and

$$\Pi_z^{(N)}(x) = (-1)^N \sum_{\substack{\omega \,\in\, \mathcal{W}(0,x) \\ |\omega| \geq 1}} z^{|\omega|} \left[\prod_{i=0}^{|\omega|} \sum_{R_i \ni \omega(i)} z^{|R_i|} \right] J^{(N)}[0,|\omega|]. \qquad (8.23)$$

By definition, $\Pi_z^{(N)}(x)$ is non-negative. By (8.8) and (8.5),

$$\Pi_z(x) = \sum_{N=1}^{\infty} (-1)^N \Pi_z^{(N)}(x). \qquad (8.24)$$

To state the diagrammatic estimates, we make the following definitions. We extract the term in the definition (7.26) of the square diagram due to $x = y = u = 0$ and define a generalized square diagram by

$$\bar{\mathsf{S}}(z) = \sup_{w \in \mathbb{Z}^d} \left[\sum_{x,y,u} G_z(x) G_z(y-x) G_z(u-y) G_z(w-u) - \delta_{0,w} g_z^4 \right]. \qquad (8.25)$$

We also define a generalized triangle diagram

$$\bar{\mathsf{T}}(z) = \sup_{u \in \mathbb{Z}^d} \left[\sum_{x,y} G_z(x) G_z(y-x) G_z(u-y) - \delta_{0,u} g_z^3 \right]. \qquad (8.26)$$

Theorem 8.2. *For $z \geq 0$ and $N \geq 1$,*

$$\sum_{x \in \mathbb{Z}^d} \Pi_z^{(N)}(x) \leq \bar{\mathsf{T}}(z) \left(2\bar{\mathsf{S}}(z) \right)^{N-1}. \qquad (8.27)$$

Proof. We discuss the proof in detail only for the case $N = 1$, and sketch the proof for $N \geq 2$.

For $N = 1$, there is a unique lace consisting of a single edge, and all other bonds are compatible with this lace. Therefore, by (8.22),

$$J^{(1)}[a,b] = U_{ab} \prod_{\substack{a \leq s < t \leq b \\ (s,t) \neq (a,b)}} (1 + U_{st}), \qquad (8.28)$$

and (8.23) then gives

$$\Pi_z^{(1)}(x) = \sum_{\substack{\omega \in \mathcal{W}(0,x) \\ |\omega| \geq 1}} z^{|\omega|} \left[\prod_{i=0}^{|\omega|} \sum_{R_i \ni \omega(i)} z^{|R_i|} \right] (-U_{0|\omega|}) \prod_{\substack{0 \leq s < t \leq |\omega| \\ (s,t) \neq (0,|\omega|)}} (1 + U_{st}).$$
$$(8.29)$$

The factor $-U_{0|\omega|}$ gives a nonzero contribution only if R_0 and $R_{|\omega|}$ intersect, and the final product in (8.29) disallows any further rib intersections. We first consider the case $x \neq 0$. Relaxing the latter restriction somewhat and overcounting an enforcement of the former gives the upper bound

$$\Pi_z^{(1)}(x) \leq \sum_{v \in \mathbb{Z}^d} \sum_{\substack{\omega \in \mathcal{W}(0,x) \\ |\omega| \geq 1}} z^{|\omega|} \sum_{R_0 \ni 0,v} |z|^{|R_0|} \sum_{R_{|\omega|} \ni x,v} z^{|R_{|\omega|}|}$$

$$\times \left[\prod_{i=1}^{|\omega|-1} \sum_{R_i \ni \omega(i), \not\ni 0,x} z^{|R_i|} \right] \prod_{1 \leq s < t \leq |\omega|-1} (1 + U_{st}). \qquad (8.30)$$

Since

$$\sum_{R_0 \ni 0, v} z^{|R_0|} = G_z(v), \qquad \sum_{R_{|\omega|} \ni x, v} z^{|R_{|\omega|}|} = G_z(v - x), \qquad (8.31)$$

and since

$$\sum_{\substack{\omega \in \mathcal{W}(0, x) \\ |\omega| \geq 1}} z^{|\omega|} \left[\prod_{i=1}^{|\omega|-1} \sum_{R_i \ni \omega(i), \not\ni 0, x} z^{|R_i|} \right] \prod_{1 \leq s < t \leq |\omega|-1} (1 + U_{st}) \leq G_z(x),$$

$$(8.32)$$

for $x \neq 0$ we have

$$\Pi_z^{(1)}(x) \leq \sum_{v \in \mathbb{Z}^d} G_z(x) G_z(v - x) G_z(v). \qquad (8.33)$$

When $x = 0$ in (8.29), we can argue using similar ideas that

$$\Pi_z^{(1)}(0) \leq \sum_{v \in \Omega} G_z(v)^2, \qquad (8.34)$$

by neglecting the interaction between the last rib and all previous ribs. In particular, the term $v = 0$ is not present in the above sum. Since $G_z(0) = g_z \geq 1$, we can combine (8.33) and (8.34) to obtain

$$\sum_{x \in \mathbb{Z}^d} \Pi_z^{(1)}(x) \leq \sum_{x, v \in \mathbb{Z}^d} G_z(v) G_z(x - v) G_z(x) - g_z^3 \leq \bar{\mathsf{T}}(z). \qquad (8.35)$$

This gives the desired result for $N = 1$. Note that the first upper bound in (8.35) is just the $u = 0$ term of the supremum in (8.26).

A similar strategy can be used to bound $\Pi_z^{(N)}(x)$ for $N \geq 2$, and we sketch the argument. The situation for $N = 2$ is shown in Fig. 8.3. For a lace $L = (0t_1, s_2|\omega|)$ consisting of exactly two edges, there are two generic configurations possible with $\prod_{st \in L} U_{st} \neq 0$, one with $s_2 = t_1$ and one with $s_2 < t_1$.

We illustrate the method for the former case. The contribution to $\Pi_z^{(2)}$ due to laces with $s_2 = t_1$ can be written

$$\sum_{\substack{\omega \in \mathcal{W}(0, x) \\ |\omega| \geq 2}} z^{|\omega|} \left[\prod_{i=0}^{|\omega|} \sum_{R_i \ni \omega(i)} z^{|R_i|} \right] \sum_{s=1}^{|\omega|-1} U_{0s} U_{s|\omega|} \prod_{s't' \in \mathcal{C}(0s, s|\omega|)} (1 + U_{s't'}).$$

$$(8.36)$$

For $U_{0s} \neq 0$, R_0 and R_s must intersect. Let y be a vertex where R_0 and R_s intersect, and let ω_{R_s} be the backbone of R_s joining $\omega(s)$ and y. For $U_{s|\omega|} \neq 0$, $R_{|\omega|}$ must intersect R_s, and hence there must be a rib emanating from a vertex on the backbone ω_{R_s} which intersects $R_{|\omega|}$ — see Fig. 8.3. By neglecting some of the rib avoidance conditions, and by arguing in a similar fashion to

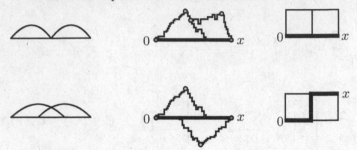

Fig. 8.3. The two generic laces consisting of two bonds, schematic diagrams showing the corresponding rib intersections for a nonzero contribution to $\Pi_z^{(2)}(x)$, and Feynman diagrams bounding the corresponding contributions to $\Pi_z^{(2)}(x)$. Diagram lines corresponding to the backbone joining 0 and x are shown in bold.

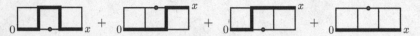

Fig. 8.4. The Feynman diagrams bounding $\Pi_z^{(N)}(x)$ for $N = 3$. Diagram lines corresponding to the backbone joining 0 and x are shown in bold.

the case $N = 1$, (8.36) can be bounded above by the Feynman diagram depicted at upper right in Fig. 8.3. The contribution from the other type of lace is bounded by the Feynman diagram at lower right in Fig. 8.3. In evaluating the diagrams, sums over vertices are constrained to disallow the coincidence of all vertices on any loop.

Explicitly, the upper bound is

$$\sum_{x\in\mathbb{Z}^d} \Pi_z^{(2)}(x) \leq 2\sum G_z(u)G_z(v)G_z(w-u)G_z(w-v)$$
$$\times G_z(x-w)G_z(y-u)G_z(x-y), \qquad (8.37)$$

where the sum is taken over all u, v, w, x, y except the terms $u = v = w = 0$ and $v = w = x = y$. We estimate this by taking the supremum over u, w of the sum over x, y of the second line, and obtain the desired upper bound $2\bar{S}(z)\bar{T}(z)$.

In general, $\Pi_z^{(N)}(x)$ is bounded above by a sum of 2^{N-1} "ladder" diagrams, each containing N non-trivial loops. The diagrams are shown in Fig. 8.4 for $N = 3$. For general N, the diagrams can be estimated inductively to prove the desired upper bound. See [95] for details. ∎

Exercise 8.3. (a) Show that the diagrams depicted in Fig. 8.4 are correct.
(b) Determine the diagrams for $N = 4$.

Percolation

This is the first of several chapters that pertain to percolation. In the present chapter, we define the model and discuss the phase transition, and then consider several differential inequalities that are relevant for the critical behaviour. The discussion is similar to that of Chap. 2 for the self-avoiding walk and the bubble condition, and to that of Chap. 7 for lattice trees and the square condition. However, now it is the triangle condition that is relevant, and the differential inequalities for percolation have a richer structure than those for the other models.

Looking ahead, in Chap. 10 we derive the lace expansion for percolation and discuss its bounds. In Chap. 11, we survey results that have been proved for percolation using the expansion. In Chaps. 12–13, we extend these ideas to oriented percolation, in which bonds are directed in the direction of increasing "time." Then in Chap. 14, we discuss applications of the expansion to the contact process, for which the time variable is continuous.

9.1 The Phase Transition

For an introduction to percolation, see [85]. Introductory material about percolation can also be found in [128], and [197] provides an introduction from a physicist's perspective.

We consider independent Bernoulli bond percolation on the integer lattice $\mathbb{Z}^d$, with edge (bond) set consisting of pairs $\{x, y\}$ of vertices of $\mathbb{Z}^d$ with $y - x \in \Omega$, where Ω defines either the nearest-neighbour model (1.1) or the spread-out model (1.2). This means that to each bond $\{x, y\}$ we associate an independent Bernoulli random variable $n_{\{x,y\}}$ which takes the value 1 with probability p and the value 0 with probability $1 - p$, where p is a parameter in the closed interval $[0, 1]$. If $n_{\{x,y\}} = 1$ then we say that the bond $\{x, y\}$ is *occupied*, and otherwise we say that it is *vacant*. A configuration is a realization of the random variables for all bonds. The joint probability distribution of the bond variables is denoted $\mathbb{P}_p$, with corresponding expectation $\mathbb{E}_p$.

Fig. 9.1. Bond percolation configurations on a 14×14 piece of the square lattice $\mathbb{Z}^2$ for $p = 0.25$, $p = 0.45$, $p = 0.55$, $p = 0.75$. The critical value is $p_c = \frac{1}{2}$.

Given a configuration and any two vertices x and y, we say that x and y are *connected*, denoted $x \leftrightarrow y$, if there is a path from x to y consisting of occupied bonds, or if $x = y$. We denote by $C(x)$ the random set of vertices connected to x, and denote its cardinality by $|C(x)|$. Let

$$\theta(p) = \mathbb{P}_p(|C(0)| = \infty). \tag{9.1}$$

The most basic fact concerning percolation is that there is a phase transition for dimensions $d \geq 2$. That is, for $d \geq 2$, there exists a critical value $p_c = p_c(d) \in (0,1)$ such that the probability $\theta(p)$ that the origin is connected to infinitely many vertices is zero for $p < p_c$, but is strictly positive for $p > p_c$. It is known that $\theta(p_c)$ is zero for certain two-dimensional models [141], and in high dimensions as discussed further in Sect. 11.1. It remains an open problem to prove that $\theta(p_c)$ is zero in general dimensions, even though it has been proved that there can be no infinite cluster in *any* half-space of $\mathbb{Z}^d$ when $p = p_c$ [85, Sect. 7.3].

The *two-point function* $\tau_p(x)$ is defined to be the probability that 0 and x are connected:

$$\tau_p(x) = \mathbb{P}_p(0 \leftrightarrow x). \tag{9.2}$$

By the translation invariance of the lattice, $\mathbb{P}_p(x \leftrightarrow y) = \tau_p(y - x)$. We will sometimes find it convenient to write the two-point function as

$$\tau_p(x, y) = \tau_p(y - x). \tag{9.3}$$

For $p < p_c$ the two-point function is known to decay exponentially as $|x| \to \infty$, so that the *correlation length*

$$\xi(p) = -\left[\lim_{n \to \infty} \frac{1}{n} \log \tau_p(ne_1) \right]^{-1}, \tag{9.4}$$

where e_1 is the unit vector $(1, \ldots, 0)$, is finite and strictly positive. The *susceptibility*, or expected cluster size, is defined by

$$\chi(p) = \sum_{x \in \mathbb{Z}^d} \tau_p(x) = \mathbb{E}_p |C(0)|. \tag{9.5}$$

An important theorem [7, 161, 162] asserts that the critical point is also characterized by

$$p_c = \sup\{p : \chi(p) < \infty\}; \tag{9.6}$$

see Exercise 9.6 below.

Let

$$P_n(p) = \mathbb{P}_p(|C(0)| = n). \tag{9.7}$$

The *magnetization* is defined, for $\gamma \in [0, 1]$, by[1]

$$M(p, \gamma) = 1 - \sum_{n=1}^{\infty} (1 - \gamma)^n P_n(p). \tag{9.8}$$

Note that

$$M(p, 0) = \theta(p). \tag{9.9}$$

The magnetization has a useful and standard probabilistic interpretation. We define i.i.d. vertex variables taking the value "green" and "not green" by declaring that each vertex is independently green with probability $\gamma \in [0, 1]$. The vertex variables are independent of the bond variables. Let $\mathcal{G}$ denote the random set of green vertices. Then, for $\gamma \neq 0$, it follows from the definition of the magnetization that

$$M(p, \gamma) = \mathbb{P}_{p,\gamma}(0 \leftrightarrow \mathcal{G}), \tag{9.10}$$

where $\{0 \leftrightarrow \mathcal{G}\}$ denotes the event that $0 \leftrightarrow x$ for some $x \in \mathcal{G}$, and $\mathbb{P}_{p,\gamma}$ denotes the joint distribution of the bond and vertex variables. Note that the limit of $M(p, \gamma)$, as $\gamma \to 0^+$, is equal to $M(p, 0) = \theta(p)$, which is not the same as $\mathbb{P}_{p,0}(0 \leftrightarrow \mathcal{G}) = 0$ for $p > p_c$.

Another quantity of interest is the *number of clusters per vertex*, defined by

$$\kappa(p) = \mathbb{E}_p(|C(0)|^{-1}) = \sum_{n=1}^{\infty} \frac{1}{n} P_n(p). \tag{9.11}$$

Various critical exponents are predicted to describe the behaviour of the above functions in the vicinity of the critical point. The critical exponents are often assumed by mathematicians to exist in a rather weak sense, which presumably should be easier to prove. However, it is worth emphasizing that the exponents are actually predicted to exist in an asymptotic sense, as indicated in the following. We use c to denote a positive constant whose value is unimportant and may change from line to line.

---

[1] This γ should not be confused with the critical exponent of (9.12), which is denoted by the same Greek letter.

On the subcritical side, the following power laws are believed to hold:

$$\chi(p) \sim c(p_c - p)^{-\gamma} \quad \text{as } p \to p_c^-, \tag{9.12}$$

$$\xi(p) \sim c(p_c - p)^{-\nu} \quad \text{as } p \to p_c^-, \tag{9.13}$$

$$\kappa'''(p) \sim c(p_c - p)^{-1-\alpha} \quad \text{as } p \to p_c^-, \tag{9.14}$$

$$\frac{\mathbb{E}_p(|C(0)|^{m+1})}{\mathbb{E}_p(|C(0)|^m)} \sim c(p_c - p)^{-\Delta} \quad \text{as } p \to p_c^-, \text{ for } m = 1, 2, \ldots, \tag{9.15}$$

for some dimension-dependent amplitudes c and critical exponents γ, ν, α, Δ. For the spread-out model, the amplitudes c may depend on L, but it is believed that the critical exponents are independent of L, as long as L is finite.

At the critical point, the following power laws are believed to hold:

$$\tau_{p_c}(x) \sim c\frac{1}{|x|^{d-2+\eta}} \quad \text{as } |x| \to \infty, \tag{9.16}$$

$$\hat{\tau}_{p_c}(k) \sim c\frac{1}{|k|^{2-\eta}} \quad \text{as } k \to 0, \tag{9.17}$$

$$M(p_c, \gamma) \sim c\gamma^{1/\delta} \quad \text{as } \gamma \to 0, \tag{9.18}$$

$$P_n(p_c) \sim c\frac{1}{n^{1+1/\delta}} \quad \text{as } n \to \infty, \tag{9.19}$$

for critical exponents η, δ. Finally, on the supercritical side, it is believed that

$$\theta(p) \sim c(p - p_c)^{\beta} \quad \text{as } p \to p_c^+, \tag{9.20}$$

for a critical exponent β. Further exponents can also be defined, e.g. the subcritical exponents have supercritical counterparts believed to take on the same values, but we restrict attention to those above, since these are the ones that will figure in the results discussed below.

Important recent advances, connected with the Schramm-Loewner evolutions, have led to proofs of existence of most critical exponents, in a weaker form than the above asymptotic form, for site percolation on the 2-dimensional triangular lattice (see [190, 191, 206] and references therein).

The lace expansion has been used to prove that the critical exponents exist and take their mean-field values in high dimensions. The mean-field values are the values assumed by the critical exponents for percolation on an infinite regular tree, namely

$$\gamma = 1, \quad \nu = \frac{1}{2}, \quad \alpha = -1, \quad \Delta = 2,$$

$$\eta = 0, \quad \delta = 2, \quad \beta = 1. \tag{9.21}$$

A derivation of these exponents for the tree can be found in [85, Sect. 10.1].

Exercise 9.1. The infinite binary tree is the infinite labelled tree in which every vertex has degree 3 except a root of degree 2. Consider bond percolation on the infinite binary tree.

(a) Let $\theta(p)$ be the probability that the root is in an infinite cluster. Note that $\eta = 1 - \theta(p)$ is the extinction probability of the Galton–Watson branching process starting from a single individual, whose offspring distribution X is binomial with parameters 2 and p. Thus η is the smallest non-negative root of $G(s) = s$, where $G(s) = \mathbb{E}s^X$ is the generating function for X (see, e.g., [87, p. 173]). Conclude that $p_c = \frac{1}{2}$ and

$$\theta(p) = \begin{cases} 0 & \text{if } p \le \frac{1}{2} \\ \frac{1}{p^2}(2p - 1) & \text{if } p \ge \frac{1}{2}, \end{cases}$$

so that $\beta = 1$.

(b) By conditioning on the number of occupied bonds containing the root, prove that $\chi(p) = (1 - 2p)^{-1}$ and hence $\gamma = 1$.

It was first predicted in [202] that the upper critical dimension for percolation is 6. The lace expansion has been applied to spread-out percolation models in dimensions $d > 6$, and to the nearest-neighbour model in higher dimensions (often $d \ge 19$ is high enough). These results show that the upper critical dimension is at most 6. On the other hand, it is known from the results of [50, 200] (see also [34]) that the upper critical dimension of percolation is at least six. In brief, these authors proved hyperscaling inequalities, such as $d\nu \ge 2\Delta - \gamma$, assuming the exponents exist. Inserting the mean-field values $\gamma = 1$, $\Delta = 2$, $\nu = \frac{1}{2}$, we see that for $d < 6$ the inequality is not satisfied and hence at least one of these exponents cannot take on its mean-field value.

9.2 Differential Inequalities

One-sided mean-field bounds for the critical exponents γ, β and δ, valid in all dimensions $d \ge 2$ can be derived from differential inequalities for the susceptibility and the magnetization. The first of these differential inequalities, for χ, is summarized in the following proposition, which is due to [12]. The proof of the proposition is deferred to Sect. 9.4.

Proposition 9.2. *For $0 < p < p_c$, and for all dimensions $d \ge 2$,*

$$\frac{d\chi(p)}{dp} \le |\Omega|\chi(p)^2. \tag{9.22}$$

In addition, $\lim_{p \to p_c^-} \chi(p) = \infty$.

The following corollary to the proposition can be interpreted as the mean-field bound $\gamma \ge 1$, valid in all dimensions $d \ge 2$.

Corollary 9.3. *For $0 < p < p_c$, and for all dimensions $d \ge 2$,*

$$\chi(p) \ge \frac{1}{|\Omega|(p_c - p)}. \tag{9.23}$$

Proof. The inequality (9.22) can be rewritten as

$$-\frac{\mathrm{d}\chi(p)^{-1}}{\mathrm{d}p} \leq |\Omega|. \tag{9.24}$$

Integration over the interval $[p, p_2]$ then gives

$$\chi(p)^{-1} - \chi(p_2)^{-1} \leq |\Omega|(p_2 - p). \tag{9.25}$$

Taking the limit $p_2 \to p_c^-$ and applying the last statement in the proposition proves the corollary. ∎

Partial differential inequalities for the magnetization are given in the following proposition, which is due to [7].

Proposition 9.4. *For $0 < p < 1$ and $0 < \gamma < 1$, and for all dimensions $d \geq 2$,*

$$(1-p)\frac{\partial M}{\partial p} \leq |\Omega|(1-\gamma)M\frac{\partial M}{\partial \gamma}, \tag{9.26}$$

$$M \leq \gamma\frac{\partial M}{\partial \gamma} + M^2 + pM\frac{\partial M}{\partial p}. \tag{9.27}$$

The derivative of M with respect to p can be eliminated from the above pair of inequalities, by inserting (9.26) into (9.27) to obtain

$$M \leq \gamma\frac{\partial M}{\partial \gamma} + M^2 + \frac{p|\Omega|}{1-p}(1-\gamma)M^2\frac{\partial M}{\partial \gamma}. \tag{9.28}$$

We omit the proof of Proposition 9.4, which can be found in [7] or [85, Sect. 5.3]. Instead, we will show how (9.26)–(9.27) can be integrated to prove the following theorem from [7]. For its statement, we define

$$\chi(p,\gamma) = (1-\gamma)\frac{\partial M(p,\gamma)}{\partial \gamma}$$

$$= \sum_{n=1}^{\infty} n(1-\gamma)^n P_n(p) \tag{9.29}$$

and

$$\chi^f(p) = \mathbb{E}_p\left[|C(0)|I[|C(0)| < \infty]\right]$$

$$= \sum_{n=1}^{\infty} nP_n(p) = \chi(p,0) \tag{9.30}$$

(the superscript f stands for *finite*).

Theorem 9.5. *Fix p with $\chi^f(p) = \infty$. There is a positive constant $a = a(p)$ such that*

$$M(p,\gamma) \geq a\gamma^{1/2}. \tag{9.31}$$

In addition, either $\theta(p) > 0$, or it is the case that $\theta(p) = 0$ and $\theta(p') \geq \frac{1}{2p'}(p' - p)$ for all $p' \geq p$.

Exercise 9.6. Define $p_c = \inf\{p : \theta(p) > 0\}$ and $\pi_c = \sup\{p : \chi(p) < \infty\}$.
(a) Prove that $\pi_c \leq p_c$.
(b) Prove that $\pi_c = p_c$. Hint: assume that $\pi_c < p_c$ and derive a contradiction from Theorem 9.5.

The bounds of Theorem 9.5 can be interpreted as the mean-field bounds $\delta \geq 2$ and $\beta \leq 1$. In fact, if $M(p_c, 0) = \theta(p_c) > 0$ then certainly $M(p_c, \gamma) \geq \theta(p_c) \geq \theta(p_c)\gamma^{1/2}$. On the other hand, if $\theta(p_c) = 0$ then $\chi^f(p_c) = \chi(p_c) = \infty$ (by Corollary 9.3), and hence

$$M(p_c, \gamma) \geq a\gamma^{1/2} \tag{9.32}$$

by Theorem 9.5. This says that $\delta \geq 2$. Similarly, if $\theta(p_c) = 0$ then Theorem 9.5 implies that

$$\theta(p) \geq \frac{1}{2p}(p - p_c) \tag{9.33}$$

for $p \geq p_c$, so that $\beta \leq 1$. On the other hand, if $\theta(p_c) > 0$ then we similarly have $\theta(p) \geq \theta(p_c) \geq \theta(p_c)(p - p_c)$, for $p \geq p_c$.

Proof of Theorem 9.5. We fix p with $\chi^f(p) = \infty$, and drop the p dependence from the notation. We first prove (9.31). Note that (9.31) is vacuous if $M(p, 0) > 0$, so we may assume that $M(p, 0) = 0$.

In (9.28), we write $A = \frac{p|\Omega|}{1-p}$, and use $1 - \gamma \leq 1$, to obtain

$$M \leq \gamma \frac{dM}{d\gamma} + M^2 + AM^2 \frac{dM}{d\gamma}. \tag{9.34}$$

We have used ordinary, rather than partial derivatives, in (9.34), since we are considering p to be fixed. We use the fact that $M > 0$ when $\gamma > 0$, and that M has a well-defined inverse function $\gamma = \gamma(M)$. Multiplication of (9.34) by $\frac{1}{M}\frac{d\gamma}{dM}$ gives

$$\frac{d\gamma}{dM} \leq \frac{\gamma}{M} + M\frac{d\gamma}{dM} + AM. \tag{9.35}$$

Equivalently,

$$\frac{d\gamma}{dM} - \frac{\gamma}{M(1-M)} \leq A\frac{M}{1-M}. \tag{9.36}$$

We multiply (9.36) by the integrating factor $\exp[-\int \frac{1}{M(1-M)}dM] = \frac{1-M}{M}$ to obtain

$$\frac{d}{dM}\left[\frac{\gamma(1-M)}{M}\right] \leq A. \tag{9.37}$$

Integration over the interval $[0, M]$ then gives

$$\frac{\gamma(1-M)}{M} - \lim_{M \to 0}\frac{\gamma(1-M)}{M} \leq AM. \tag{9.38}$$

To evaluate the limit, we observe that $\gamma(0) = 0$ (recall that we are assuming $M(p, 0) = 0$), and conclude from the mean-value theorem and (9.29) that there is a $\gamma^* \in (0, \gamma)$ such that

$$\frac{M}{\gamma} = \frac{dM}{d\gamma}\Big|_{\gamma=\gamma^*} = \frac{1}{1-\gamma^*}\chi(p, \gamma^*) \to \chi^f(p) = \infty \qquad (9.39)$$

as $\gamma \to 0$. Therefore,

$$\frac{\gamma(1 - M)}{M} \leq AM. \qquad (9.40)$$

Since $M \leq \frac{1}{2}$ for small γ (because $M \to 0$ as $\gamma \to 0$), it follows that

$$M^2 \geq \frac{1}{2A}\gamma \qquad (9.41)$$

for small γ, which proves (9.31).

Next, we consider the statement for θ. If $\theta(p) > 0$ then we are done, so we assume $\theta(p) = 0$. We multiply (9.27) by $\frac{1}{\gamma M}$ to obtain

$$0 \leq \frac{\partial \log M}{\partial \gamma} + \frac{1}{\gamma}\frac{\partial}{\partial \tilde{p}}(\tilde{p}M - \tilde{p}), \qquad (9.42)$$

where $\tilde{p} \in (0, 1)$ is arbitrary (and M is evaluated at $\tilde{p}$). We fix $0 < \gamma_1 < \gamma_2 < 1$ and integrate (9.42) with respect to γ over the interval $[\gamma_1, \gamma_2]$, and with respect to $\tilde{p}$ over the interval $[p, p']$.

For the first term, we integrate first with respect to γ, and use the monotonicity of M, to obtain

$$\log M(\tilde{p}, \gamma_2) - \log M(\tilde{p}, \gamma_1) \leq \log M(p', \gamma_2) - \log M(p, \gamma_1), \qquad (9.43)$$

obtaining

$$(p' - p)[\log M(p', \gamma_2) - \log M(p, \gamma_1)] \qquad (9.44)$$

as an upper bound for the double integral of the first term.

For the second term, we integrate first with respect to $\tilde{p}$, using

$$p'M(p', \gamma) - pM(p, \gamma) - (p' - p) \leq p'M(p', \gamma_2) - p' + p, \qquad (9.45)$$

and then complete the integration to obtain the upper bound

$$[p'M(p', \gamma_2) - p' + p] \log \frac{\gamma_2}{\gamma_1} \qquad (9.46)$$

for the double integral of the second term.

Altogether, this gives

$$0 \leq (p' - p)\left[\frac{\log M(p', \gamma_2)}{\log \frac{\gamma_2}{\gamma_1}} - \frac{\log M(p, \gamma_1)}{\log \frac{\gamma_2}{\gamma_1}}\right] + p'M(p', \gamma_2) - p' + p. \qquad (9.47)$$

Now we let $\gamma_1 \to 0$. The first term vanishes in this limit. For the second term, we use (9.31) to conclude that

$$-\frac{\log M(p, \gamma_1)}{\log \frac{\gamma_2}{\gamma_1}} \leq \frac{-\log a - \frac{1}{2}\log \gamma_1}{\log \gamma_2 - \log \gamma_1}. \tag{9.48}$$

Since the right hand side approaches $\frac{1}{2}$ as $\gamma_1 \to 0$, we obtain

$$0 \leq \frac{1}{2}(p' - p) + p'M(p', \gamma_2) - p' + p, \tag{9.49}$$

which implies that

$$M(p', \gamma_2) \geq \frac{1}{2p'}(p' - p). \tag{9.50}$$

Now we take the limit $\gamma_2 \to 0$ to complete the proof. ∎

9.3 Differential Inequalities and the Triangle Condition

In this section, we state differential inequalities involving the triangle diagram which are complementary to the differential inequalities (9.22) for χ and (9.28) for M. We also derive consequences of these differential inequalities.

Let

$$\nabla_p(w) = \sum_{x,y \in \mathbb{Z}^d} \tau_p(x)\tau_p(y - x)\tau_p(w - y). \tag{9.51}$$

By Exercise 1.1,

$$\hat{\nabla}_p(k) = \hat{\tau}_p(k)^3. \tag{9.52}$$

The *triangle diagram* is defined by

$$\mathsf{T}(p) = \nabla_p(0) = \int_{[-\pi,\pi]^d} \hat{\tau}_p(k)^3 \frac{d^dk}{(2\pi)^d}, \tag{9.53}$$

and the *triangle condition* is the statement that $\mathsf{T}(p_c) < \infty$. We also define

$$\overline{\mathsf{T}}(p) = \sup_{w \in \mathbb{Z}^d} [\nabla_p(w) - \delta_{w,0}]. \tag{9.54}$$

A proof that $\hat{\tau}_p(k) \geq 0$ is given in [12], so the integrand of (9.53) is non-negative. If we insert the conjectured behaviour $\hat{\tau}_{p_c}(k) \sim c|k|^{\eta-2}$ into the integral of (9.53), the result is finite if $d > 6 - 3\eta$, and infinite otherwise. The mean-field value of η is zero, $\eta = \frac{5}{24}$ for $d = 2$ (this has currently been proven only for site percolation on the triangular lattice [191]), and η is conjectured to be negative for $2 < d < 6$ [1]. Thus, the triangle condition should hold if and only if $d > 6$.

The following differential inequality is complementary to (9.22). We defer its proof to Sect. 9.4.

Proposition 9.7. *For $0 < p < p_c$, and for all dimensions $d \geq 2$,*

$$\frac{d\chi(p)}{dp} \geq [1 - \bar{\mathsf{T}}(p)]|\Omega|\chi(p)^2. \tag{9.55}$$

The following corollary to the proposition shows that $\gamma \leq 1$ if $\bar{\mathsf{T}}(p_c) < 1$. Since we have seen quite generally that $\gamma \geq 1$, this means that $\gamma = 1$ if $\bar{\mathsf{T}}(p_c) < 1$. In fact, it can be shown that $\gamma \leq 1$ whenever the triangle condition $\mathsf{T}(p_c) < \infty$ holds [12], rather than the stronger assumption $\bar{\mathsf{T}}(p_c) < 1$. However, in practice, the stronger conclusion $\bar{\mathsf{T}}(p_c) < 1$ is what has been proved using the lace expansion, and this will suffice for our purposes. (As in the discussion of Sect. 5.3, it would be desirable to find an approach to the lace expansion that is based on the finiteness of the triangle diagram, rather than on its smallness.)

Corollary 9.8. *Let $d \geq 2$ and $0 \leq p < p_c$. If $\bar{\mathsf{T}}(p_c) < 1$ then*

$$\chi(p) \leq \frac{1}{[1 - \bar{\mathsf{T}}(p_c)]|\Omega|(p_c - p)}. \tag{9.56}$$

Proof. By monotonicity, $\bar{\mathsf{T}}(p)$ can be replaced by $\bar{\mathsf{T}}(p_c)$ in (9.55). As in the proof of Corollary 9.3, integration of the resulting inequality over (p_1, p_2) (with $p_2 < p_c$) gives

$$\chi(p_1)^{-1} - \chi(p_2)^{-1} \geq [1 - \bar{\mathsf{T}}(p_c)]|\Omega|(p_2 - p_1). \tag{9.57}$$

Taking $p_1 = p$ and $p_2 \to p_c^-$, the desired result then follows from the fact that $\chi(p) \to \infty$ as $p \to p_c^-$, by Proposition 9.2. ∎

For the magnetization, we have the following differential inequality involving the triangle. This differential inequality is a variant of an inequality derived by [20], and is complementary to (9.28). It was proved in essentially this form in [31]. The proof will be given in Sect. 9.4. A factor M can be cancelled on each side of (9.58), but we state the inequality in this form because this is how it is proved, and for comparison with (9.28).

Proposition 9.9. *Let $0 < \gamma < 1$ and $p \in [|\Omega|^{-1}, p_c)$. There is a universal constant a_0 such that if $\bar{\mathsf{T}}(p_c)$ is sufficiently small then*

$$M \geq a_0 p|\Omega|(1 - \gamma)M^2 \frac{\partial M}{\partial \gamma}. \tag{9.58}$$

The following theorem is a consequence of Proposition 9.9. When combined with Theorem 9.5, it shows that $M(p_c, \gamma) \simeq \gamma^{1/2}$ and $\theta(p) \simeq (p_c - p)$, i.e., that the critical exponents δ and β exist and take their mean-field values $\delta = 2$ and $\beta = 1$, whenever $\bar{\mathsf{T}}(p_c)$ is sufficiently small. This will be sufficient for our needs, but it should be noted that in [20] the same conclusion is derived under the weaker assumption that $\mathsf{T}(p_c)$ is finite. Our proof of Theorem 9.10 is based on [31].

Theorem 9.10. *Let $a_1^2 = \max\{2, 4/a_0\}$, where a_0 is the constant of Proposition 9.9. If $\bar{\mathsf{T}}(p_c)$ is sufficiently small, then for $\gamma \geq 0$ and $p \geq p_c$,*

$$M(p_c, \gamma) \leq a_1 \gamma^{1/2}, \tag{9.59}$$
$$\theta(p) \leq 4a_1^2 |\Omega|(p - p_c). \tag{9.60}$$

In particular, $\theta(p_c) = 0$.

Proof. We begin with the proof of the upper bound on $M(p_c, \gamma)$. Suppose first that $\gamma \leq \frac{1}{2}$, so that $1 - \gamma \geq \frac{1}{2}$. Let $p \in [|\Omega|^{-1}, p_c)$ (it is a standard fact that $p_c > |\Omega|^{-1}$; see [85, (1.13)]). Then $p|\Omega| \geq 1$ and it follows from (9.58) that

$$\frac{\partial M^2}{\partial \gamma} \leq \frac{4}{a_0} \leq a_1^2. \tag{9.61}$$

Therefore,

$$M(p, \gamma)^2 - M(p, 0)^2 \leq a_1^2 \gamma. \tag{9.62}$$

Since $p < p_c$, the subtracted term on the right hand side is zero. Taking the limit $p \to p_c^-$, and using the continuity of $M(p, \gamma)$ in p for $\gamma > 0$, we obtain the desired bound

$$M(p_c, \gamma) \leq a_1 \gamma^{1/2} \tag{9.63}$$

for $\gamma > 0$. The same bound then holds for $\gamma = 0$ due to the monotonicity of $M(p_c, \gamma)$ in γ. On the other hand, if $\gamma \geq \frac{1}{2}$, then

$$M(p_c, \gamma)^2 \leq 1 \leq 2\gamma, \tag{9.64}$$

which gives (9.59) in this case.

To extend (9.59) to (9.60), we apply an *extrapolation principle* [7,9,20,73]. The extrapolation principle converts the upper bound (9.59) on $M(p_c, \gamma)$ to an upper bound valid for $p > p_c$. This is perhaps surprising, since M is an increasing function of p. However, it is also increasing in γ, and the differential inequality (9.26) can be used to compensate for an increase in p with a decrease in γ.

In our context of nearest-neighbour or spread-out bond percolation on $\mathbb{Z}^d$, we have $p_c \leq \frac{1}{2}$. Suppose first that $p > \frac{3}{4}$. Then

$$\theta(p) \leq 1 \leq 4(p - p_c) \leq 4a_1^2 |\Omega|(p - p_c). \tag{9.65}$$

Thus we may assume that $p \in (p_c, \frac{3}{4})$, and we make this assumption throughout the rest of the proof.

We find it convenient to make the change of variables $\gamma = 1 - e^{-h}$, and define $\tilde{M}(p, h) \doteq M(p, 1 - e^{-h})$, for $h \geq 0$. Since $p \leq \frac{3}{4}$, the differential inequality (9.26) implies that

$$\frac{\partial \tilde{M}}{\partial p} \leq 4|\Omega|\tilde{M}\frac{\partial \tilde{M}}{\partial h}. \tag{9.66}$$

For fixed fixed $p \in (0,1)$ and $m \in [\theta(p),1)$, we can solve the equation $\tilde{M}(p,h) = m$ for $h = h(p)$, so that $\tilde{M}(p, h(p)) = m$. Differentiation of this identity with respect to p gives

$$\frac{\partial \tilde{M}}{\partial p} + \frac{\partial \tilde{M}}{\partial h} \frac{\partial h}{\partial p}\bigg|_{\tilde{M}=m} = 0. \tag{9.67}$$

Therefore,

$$0 \leq -\frac{\partial h}{\partial p}\bigg|_{\tilde{M}=m} = \frac{\frac{\partial \tilde{M}}{\partial p}}{\frac{\partial \tilde{M}}{\partial h}} \leq 4|\Omega|m. \tag{9.68}$$

Choose $p \in (p_c, \frac{3}{4})$, let P_1 be the point $(p,0)$, and set $m_1 = \tilde{M}(p,0)$. The bound $-\frac{\partial h}{\partial p}\big|_{\tilde{M}=m_1} \leq 4|\Omega|m_1$ implies that the point $P_2 = (p_c, 4|\Omega|m_1(p-p_c))$ lies above the contour line $\tilde{M} = m_1$ in the (p,h) plane (see Fig. 9.2). Since $\tilde{M}(p_c,h)$ is monotone in h, the value of $\tilde{M}$ at P_2 is bounded below by the value at P_1.

Therefore,

$$\theta(p) = M(p,0) \leq M(p_c, 4|\Omega|\theta(p)(p-p_c)). \tag{9.69}$$

Applying (9.59) gives

$$\theta(p) \leq a_1\sqrt{4|\Omega|\theta(p)(p-p_c)}, \tag{9.70}$$

which implies that

$$\theta(p) \leq 4a_1^2|\Omega|(p-p_c). \tag{9.71}$$

This completes the proof.   ∎

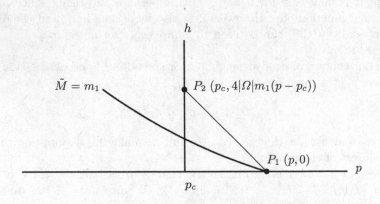

Fig. 9.2. The extrapolation geometry.

9.4 Proofs of the Differential Inequalities

9.4.1 Differential Inequality for the Susceptibility

In this section, we sketch the proof of the differential inequalities for the susceptibility given in Propositions 9.2 and 9.7, which are restated here as Proposition 9.11. We follow the original reasoning of Aizenman and Newman [12].

Proposition 9.11. *For $d \geq 2$ and $0 < p < p_c$,*

$$[1 - \bar{\mathsf{T}}(p)]|\Omega|\chi(p)^2 \leq \frac{\mathrm{d}\chi(p)}{\mathrm{d}p} \leq |\Omega|\chi(p)^2. \tag{9.72}$$

In addition, $\lim_{p \to p_c^-} \chi(p) = \infty$.

A complete proof requires an approximation of $\mathbb{Z}^d$ by a finite graph, followed by a limiting argument. This is quite similar to the argument used in the proof of Theorem 7.6, and we will omit the details here. Full details can be found in [12]. The proof that χ diverges as p approaches p_c^- uses the finite volume argument, and will also be omitted here.

We write $\mathbb{B}$ for the set of all bonds. Given a bond configuration, a bond $\{u, v\} \in \mathbb{B}$ (occupied or not) is called *pivotal* for the connection from x to y if $x \leftrightarrow y$ in the possibly modified configuration in which the bond is made occupied, whereas x is not connected to y in the possibly modified configuration in which the bond is made vacant. Bonds are not usually regarded as directed. However, it will be convenient at times to regard a bond $\{u, v\}$ as directed from u to v, and we will emphasize this point of view with the notation (u, v). A directed bond (u, v) is pivotal for the connection from x to y if $x \leftrightarrow u$, if $v \leftrightarrow y$, and if x is not connected to y if $\{u, v\}$ is made vacant.

Let $E \circ F$ denote the event that the events E and F occur disjointly. In our applications, E and F will be events that certain connections take place, and in this case $E \circ F$ is the event that both E and F occur and that given a configuration it is possible to find one set of bonds providing the connections needed by E and a disjoint set of bonds providing the connections needed by F. The BK inequality implies that for such events E and F, $\mathbb{P}_p(E \circ F) \leq \mathbb{P}_p(E)\mathbb{P}_p(F)$ (see [85, Theorem 2.12] for a more complete and precise discussion). The BK inequality has great importance in percolation theory, and will be an key ingredient in our analysis. We will use it in the form of the following proposition. The proposition plays the role for percolation played by subadditive estimates for the self-avoiding walk and lattice trees, and expresses a repulsive character of the model.

Proposition 9.12. *Let $V_1, V_2, \ldots, V_n$ be sets of paths in the lattice, and let E_i $(i = 1, \ldots, n)$ be the event that at least one of the paths in V_i is occupied. Let F be the event that there are pairwise edge-disjoint occupied paths from each of the sets $V_1, V_2, \ldots, V_n$. Then*

$$\mathbb{P}_p(F) \leq \prod_{i=1}^{n} \mathbb{P}_p(E_i).$$

Sketch of proof of the upper bound in (9.72). By Russo's formula (see [85, Theorem 2.25]),

$$\frac{\mathrm{d}}{\mathrm{d}p} \tau_p(x, y) = \sum_{\{u,v\} \in \mathbb{B}} \mathbb{P}_p(\{u, v\} \text{ is pivotal for } x \leftrightarrow y)$$

$$= \sum_{(u,v)} \mathbb{P}_p((u, v) \text{ is pivotal for } x \leftrightarrow y), \qquad (9.73)$$

where the sum over (u, v) is a sum over *directed* bonds. The above is fine for a finite graph, but is not directly legitimate for an infinite graph. We will ignore this subtlety. By (9.73) and the BK inequality,

$$\frac{\mathrm{d}}{\mathrm{d}p} \tau_p(x, y) \leq \sum_{(u,v)} \mathbb{P}_p(\{x \leftrightarrow u\} \circ \{v \leftrightarrow y\}) \leq \sum_{(u,v)} \tau_p(x, u) \tau_p(v, y). \qquad (9.74)$$

We then perform the sums over y, v, u (in that order) and use translation invariance to obtain the desired upper bound. ∎

For the lower bound of (9.72), we will use the following definition.

Definition 9.13. *(a) Given a bond configuration, and a random or deterministic set $A \subset \mathbb{Z}^d$, we say x and y are* connected in A *if there is an occupied path from x to y whose vertices are all in A, or if $x = y \in A$. We define a* restricted two-point function *by*

$$\tau^A(x, y) = \mathbb{P}_p(x \leftrightarrow y \text{ in } \mathbb{Z}^d \backslash A). \qquad (9.75)$$

(b) Given a bond configuration, and a random or deterministic set $A \subset \mathbb{Z}^d$, we say x and y are connected through A, *denoted $x \overset{A}{\leftrightarrow} y$, if $x = y \in A$ or if $x \leftrightarrow y$ and every occupied path connecting x to y has at least one bond with an endpoint in A, i.e., if x is connected to y but it is not the case that $x \leftrightarrow y$ in $\mathbb{Z}^d \backslash A$.*

(c) Given a bond configuration, and a bond b, we define $\tilde{C}^b(x)$ to be the set of vertices connected to x in the new configuration obtained by setting b to be vacant.

Exercise 9.14. Prove that

$$\{(u, v) \text{ pivotal for } 0 \leftrightarrow x\}$$
$$= \{0 \leftrightarrow u \text{ in } \tilde{C}^{(u,v)}(0)\} \cap \{v \leftrightarrow x \text{ in } \mathbb{Z}^d \backslash \tilde{C}^{(u,v)}(0)\}. \qquad (9.76)$$

Sketch of proof of the lower bound in (9.72). Conditioning on $\tilde{C}^{(u,v)}(0)$, it follows from (9.76) that

$$\mathbb{P}_p((u,v) \text{ pivotal for } 0 \leftrightarrow x)$$

$$= \sum_{A:A \ni 0} \mathbb{P}_p(\tilde{C}^{(u,v)}(0) = A, 0 \leftrightarrow u \text{ in } A, v \leftrightarrow x \text{ in } \mathbb{Z}^d \backslash A)$$

$$= \sum_{A:A \ni 0} \mathbb{P}_p(\tilde{C}^{(u,v)}(0) = A, 0 \leftrightarrow u \text{ in } A)\mathbb{P}_p(v \leftrightarrow x \text{ in } \mathbb{Z}^d \backslash A)$$

$$= \sum_{A:A \ni 0} \mathbb{P}_p(\tilde{C}^{(u,v)}(0) = A, 0 \leftrightarrow u)\tau_p^A(v,x). \qquad (9.77)$$

Here, the sum over A is the sum over lattice animals containing 0. Also, we have used the fact that the events within the probabilities in the third line are independent, and, in the last line, that if $0 \leftrightarrow u$ but it is not the case that $0 \leftrightarrow u$ in A then v must be in A and hence $\tau_p^A(v,x) = 0$. It follows that

$$\mathbb{P}_p((u,v) \text{ pivotal for } 0 \leftrightarrow x) = \mathbb{E}_p\left(I[0 \leftrightarrow u]\tau^{\tilde{C}^{(u,v)}(0)}(v,x)\right). \qquad (9.78)$$

In the nested expectation on the right hand side of (9.78), the set $\tilde{C}^{(u,v)}(w)$ is a random set with respect to the expectation $\mathbb{E}_p$, but it is deterministic with respect to the expectation defining the restricted two-point function. The latter expectation effectively introduces a second percolation model on a second lattice, which is coupled to the original percolation model via the set $\tilde{C}^{(u,v)}(0)$.

Since

$$\tau_p^A(v,x) = \tau_p(v,x) - \left[\tau_p(v,x) - \tau_p^A(v,x)\right]$$

$$= \tau_p(v,x) - \mathbb{P}_p(v \overset{A}{\leftrightarrow} x), \qquad (9.79)$$

the identity (9.78) can be rewritten as

$$\mathbb{P}_p((u,v) \text{ pivotal for } 0 \leftrightarrow x)$$

$$= \tau_p(0,u)\tau_p(v,x) - \mathbb{E}_p\left(I[0 \leftrightarrow u]\mathbb{P}_p(v \xleftarrow{\tilde{C}^{(u,v)}(0)} x)\right). \qquad (9.80)$$

By the BK inequality, for $A \subset \mathbb{Z}^d$ we have

$$\mathbb{P}_p(v \overset{A}{\leftrightarrow} x) \leq \mathbb{P}_p\left(\bigcup_{y \in A}\{v \leftrightarrow y\} \circ \{y \leftrightarrow x\}\right)$$

$$\leq \sum_{y \in \mathbb{Z}^d} I[y \in A]\mathbb{P}_p(\{v \leftrightarrow y\} \circ \{y \leftrightarrow x\})$$

$$\leq \sum_{y \in \mathbb{Z}^d} I[y \in A]\tau_p(v,y)\tau_p(y,x). \qquad (9.81)$$

Therefore, for $A = \tilde{C}^{(u,v)}(0) \subset C(0)$, we have

$$\mathbb{P}_p(v \xleftarrow{\tilde{C}^{(u,v)}(0)} x) \leq \sum_{y \in \mathbb{Z}^d} I[y \in C(0)] \tau_p(v,y) \tau_p(y,x). \tag{9.82}$$

Substitution yields

$$\mathbb{P}_p((u,v) \text{ pivotal for } 0 \leftrightarrow x)$$
$$\geq \tau_p(0,u)\tau_p(v,x) - \sum_{y \in \mathbb{Z}^d} \mathbb{P}_p(0 \leftrightarrow u, 0 \leftrightarrow y)\tau_p(v,y)\tau_p(y,x). \tag{9.83}$$

The tree-graph bound [12], which is an elementary consequence of the BK inequality, implies that

$$\mathbb{P}_p(0 \leftrightarrow u, 0 \leftrightarrow y) \leq \sum_{z \in \mathbb{Z}^d} \tau_p(0,z)\tau_p(z,y)\tau_p(z,u). \tag{9.84}$$

Therefore,

$$\mathbb{P}_p((u,v) \text{ pivotal for } 0 \leftrightarrow x) \geq \tau_p(0,u)\tau_p(v,x) \tag{9.85}$$
$$- \sum_{y,z \in \mathbb{Z}^d} \tau_p(0,z)\tau_p(z,y)\tau_p(z,u)\tau_p(y,v)\tau_p(y,x).$$

See Fig. 9.3. Recalling (9.73) (again we are careless about the infinite volume limit), and performing the sums over u, v, x, leads to

$$\frac{d\chi(p)}{dp} \geq |\Omega|\chi(p)^2 - \chi(p) \sum_{z \in \mathbb{Z}^d} \tau_p(0,z) \sum_{(u,v)} \sum_{y \in \mathbb{Z}^d} \tau_p(z,y)\tau_p(z,u)\tau_p(y,v)$$
$$= |\Omega|\chi(p)^2 - \chi(p) \sum_{z \in \mathbb{Z}^d} \tau_p(0,z)|\Omega| \sup_{e \in \Omega} \nabla_p(e)$$
$$\geq |\Omega|\chi(p)^2[1 - \bar{\mathsf{T}}(p)], \tag{9.86}$$

using symmetry in the second step. ∎

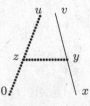

Fig. 9.3. The subtracted term in (9.85). The dotted lines arise from paths in $\tilde{C}^{\{u,v\}}(0)$, while the solid line arises from a connection from v to x through $\tilde{C}^{\{u,v\}}(0)$.

9.4.2 Differential Inequality for the Magnetization

In this section, we prove the differential inequality (9.58), following the method of [31]. Our proof is related to, but simpler than, the method used in [20] to prove an inequality useful under the weaker assumption that the tri-angle condition holds (rather than that $\bar{\mathsf{T}}(p_c)$ is small). Nevertheless, the proof is not simple. Results related to (9.58) can also be found in [104, Sect. 3].

We restate (9.58) as (9.87) in the following proposition. Note that, by (9.29), the factor $(1 - \gamma)\partial M/\partial \gamma$ on the right hand side of (9.58) can be replaced by $\chi(p, \gamma)$.

Proposition 9.15. *Let $0 < \gamma < 1$ and $p \in [|\Omega|^{-1}, p_c)$. There is a universal constant a_0 such that if $\bar{\mathsf{T}}(p_c)$ is sufficiently small then*

$$M \geq a_0 p |\Omega| M^2 \chi(p, \gamma). \tag{9.87}$$

Before beginning the proof, we discuss some preliminaries. Recall from (9.10) that $M(p, \gamma) = \mathbb{P}_{p,\gamma}(0 \leftrightarrow \mathcal{G})$. We define $\{x \leftrightarrow \mathcal{G}\}$ to be the event that $C(x) \cap \mathcal{G} \neq \varnothing$. We say that (u, v) is pivotal for $x \leftrightarrow \mathcal{G}$ if $x \leftrightarrow u$, if $v \leftrightarrow \mathcal{G}$, and if $x \nleftrightarrow \mathcal{G}$ when (u, v) is made vacant. Let $\{v \Leftrightarrow \mathcal{G}\}$ denote the event that there exist $x, y \in \mathcal{G}$, with $x \neq y$, such that there are disjoint connections $v \leftrightarrow x$ and $v \leftrightarrow y$. Let $F_{(u,v)}$ denote the event that (i) the bond (u, v) is occupied and pivotal for the connection from 0 to $\mathcal{G}$, and that (ii) $\{v \Leftrightarrow \mathcal{G}\}$. Let $F = \cup_{(u,v)} F_{(u,v)}$, and note that the union is disjoint. Since $0 \leftrightarrow \mathcal{G}$ when F occurs,

$$M = \mathbb{P}_{p,\gamma}(0 \leftrightarrow \mathcal{G}) \geq \mathbb{P}_{p,\gamma}(F) = \sum_{(u,v)} \mathbb{P}_{p,\gamma}(F_{(u,v)}), \tag{9.88}$$

and it suffices to prove that $\mathbb{P}_{p,\gamma}(F)$ is bounded below by the right hand side of (9.87).

Let $0 < \gamma < 1$. For $x, y \in \mathbb{V}$, we define a "green-free" analogue of the two-point function by

$$\tau_{p,\gamma}(x, y) = \mathbb{P}_{p,\gamma}(x \leftrightarrow y, x \nleftrightarrow \mathcal{G}), \tag{9.89}$$

so that

$$\chi(p, \gamma) = \sum_{x \in \mathbb{Z}^d} \tau_{p,\gamma}(0, x). \tag{9.90}$$

Exercise 9.16. Prove (9.90). (Recall (9.29).)

Given a subset $A \subset \mathbb{V}$, we define $\tau_{p,\gamma}^A(x, y)$ to be the probability that (i) $x \leftrightarrow y$ in $\mathbb{Z}^d \backslash A$, and (ii) $x \nleftrightarrow \mathcal{G}$ in $\mathbb{Z}^d \backslash A$, which is to say that $x \nleftrightarrow \mathcal{G}$ *after* every bond with an endpoint in A is made vacant. We write $\tilde{I}^{\{u,v\}}[E]$ for the indicator that E occurs after $\{u, v\}$ is made vacant.

Lemma 9.17.

$$\mathbb{P}_{p,\gamma}(F_{(u,v)}) = p\mathbb{E}_{p,\gamma}\left[\tau_{p,\gamma}^{\tilde{C}^{(u,v)}(v)}(0,u)\tilde{I}^{\{u,v\}}[v \Leftrightarrow \mathcal{G}]\right]. \qquad (9.91)$$

Proof. We first observe that the event $F_{(u,v)}$ is equal to

$$\{0 \leftrightarrow u \text{ in } \mathbb{Z}^d \backslash \tilde{C}^{(u,v)}(v)\} \cap \{0 \nleftrightarrow \mathcal{G} \text{ in } \mathbb{Z}^d \backslash \tilde{C}^{(u,v)}(v)\}$$
$$\cap \{\{u,v\} \text{ occupied}\}$$
$$\cap \{v \Leftrightarrow \mathcal{G} \text{ after } \{u,v\} \text{ made vacant}\}. \qquad (9.92)$$

Exercise 9.18. Prove that $F_{(u,v)}$ is equal to the above event.

The bond $\{u,v\}$ has an endpoint in $\tilde{C}^{(u,v)}(v)$, and hence the event that $\{u,v\}$ is occupied is independent of the other events above. Therefore,

$$\mathbb{P}_{p,\gamma}(F_{(u,v)}) = p\sum_{A:A\ni v}\mathbb{E}_{p,\gamma}\left[I[\tilde{C}^{(u,v)}(v) = A]\tilde{I}^{\{u,v\}}[v \Leftrightarrow \mathcal{G}] \qquad (9.93)$$
$$\times I[(0 \leftrightarrow u \text{ and } 0 \nleftrightarrow \mathcal{G}) \text{ in } \mathbb{Z}^d \backslash A]].$$

The two events in the first line depend only on bonds with an endpoint in A (but not on $\{u,v\}$) and vertices in A, while those in the second line depend only on bonds with no endpoint in A (so not on $\{u,v\}$) and on vertices in $\mathbb{Z}^d \backslash A$. Therefore,

$$\mathbb{P}_{p,\gamma}(F_{(u,v)}) = p\sum_{A:A\ni v}\mathbb{E}_{p,\gamma}\left[I[\tilde{C}^{(u,v)}(v) = A]\tilde{I}^{\{u,v\}}[v \Leftrightarrow \mathcal{G}]\right]\tau_{p,\gamma}^A(0,u), \quad (9.94)$$

which implies the desired result. ∎

Proof of Proposition 9.15. We use the identities

$$\tau_{p,\gamma}^{\tilde{C}^{(u,v)}(v)}(0,u) = \tau_{p,\gamma}(0,u) - \left(\tau_{p,\gamma}(0,u) - \tau_{p,\gamma}^{\tilde{C}^{(u,v)}(v)}(0,u)\right) \qquad (9.95)$$

and

$$\tilde{I}^{\{u,v\}}[v \Leftrightarrow \mathcal{G}] = I[v \Leftrightarrow \mathcal{G}] - \left(I[v \Leftrightarrow \mathcal{G}] - \tilde{I}^{\{u,v\}}[v \Leftrightarrow \mathcal{G}]\right). \qquad (9.96)$$

Recalling (9.90), it follows from Lemma 9.17 that

$$\mathbb{P}_{p,\gamma}(F) = p|\Omega|\chi(p,\gamma)\mathbb{P}_{p,\gamma}(0 \Leftrightarrow \mathcal{G}) \qquad (9.97)$$
$$- p\sum_{(u,v)}\tau_{p,\gamma}(0,u)\mathbb{E}_{p,\gamma}\left[I[v \Leftrightarrow \mathcal{G}] - \tilde{I}^{\{u,v\}}[v \Leftrightarrow \mathcal{G}]\right]$$
$$- p\sum_{(u,v)}\mathbb{E}_{p,\gamma}\left[\left(\tau_{p,\gamma}(0,u) - \tau_{p,\gamma}^{\tilde{C}^{(u,v)}(v)}(0,u)\right)\tilde{I}^{\{u,v\}}[v \Leftrightarrow \mathcal{G}]\right].$$

We write (9.97) as $X_1 - X_2 - X_3$, bound X_1 from below, and bound X_2 and X_3 from above.

Lower bound on X_1. We will prove that

$$\mathbb{P}_{p,\gamma}(0 \Leftrightarrow \mathcal{G}) \geq \binom{|\Omega|}{2} p^2 (1-p)^{|\Omega|-2} M(p,\gamma)^2 \left[(1 - \bar{\mathsf{T}}(p))^2 - \bar{\mathsf{T}}(p)\right]. \quad (9.98)$$

The factor $\binom{|\Omega|}{2} p^2$ is bounded below by a universal constant, since $p|\Omega| \geq 1$. Also, setting setting $p_1 = 0$ and $p_2 = p_c$ in (9.57) gives $p_c|\Omega| \leq [1 - \bar{\mathsf{T}}(p_c)]^{-1}$, and therefore the factor $(1-p)^{|\Omega|-2}$ is bounded below by a universal constant. The final factor, in square brackets, is as close to 1 as desired, for $\bar{\mathsf{T}}(p_c)$ sufficiently small, and thus there is a universal constant a such that

$$X_1 \geq p|\Omega|\chi(p,\gamma)aM(p,\gamma)^2. \quad (9.99)$$

To prove (9.98), we first note that the event $\{0 \Leftrightarrow \mathcal{G}\}$ contains the event $\cup_{e,f} E_{e,f}$, where the union is over unordered pairs of vertices $e, f \in \Omega$, the union is disjoint, and the event $E_{e,f}$ is defined as follows. Let $E_{e,f}$ be the event that the bonds $\{0, e\}$ and $\{0, f\}$ are occupied, all other bonds incident on 0 are vacant, and that in the reduced graph $\mathbb{Z}_-^d$ obtained by deleting the origin and each of the $|\Omega|$ bonds incident on 0 from $\mathbb{Z}^d$, the following three events occur: $e \leftrightarrow \mathcal{G}$, $f \leftrightarrow \mathcal{G}$, and $C(e) \cap C(f) = \varnothing$. Let $\mathbb{P}_{p,\gamma}^-$ denote the joint bond/vertex measure on $\mathbb{Z}_-^d$. Then

$$\mathbb{P}_{p,\gamma}(0 \Leftrightarrow \mathcal{G}) \geq \mathbb{P}_{p,\gamma}\left(\cup_{e,f} E_{e,f}\right) = \sum_{e,f} \mathbb{P}_{p,\gamma}(E_{e,f}) \quad (9.100)$$

$$= p^2(1-p)^{|\Omega|-2} \sum_{e,f} \mathbb{P}_{p,\gamma}^-(e \leftrightarrow \mathcal{G}, \ f \leftrightarrow \mathcal{G}, \ C(e) \cap C(f) = \varnothing).$$

Let W denote the event whose probability appears on the right hand side of (9.100). Conditioning on the set $C(e) = A \subset \mathbb{Z}_-^d$, we see that

$$\mathbb{P}_{p,\gamma}^-(W) = \sum_{A:A \ni e} \mathbb{P}_{p,\gamma}^-(C(e) = A, \ e \leftrightarrow \mathcal{G}, \ f \leftrightarrow \mathcal{G}, \ A \cap C(f) = \varnothing). \quad (9.101)$$

The above can be rewritten as

$$\mathbb{P}_{p,\gamma}^-(W) = \sum_{A:A \ni e} \mathbb{P}_{p,\gamma}^-(C(e) = A, \ e \leftrightarrow \mathcal{G}, \ f \leftrightarrow \mathcal{G} \text{ in } \mathbb{Z}_-^d \setminus A), \quad (9.102)$$

where $\{f \leftrightarrow \mathcal{G} \text{ in } \mathbb{Z}_-^d \setminus A\}$ is the event that there exists $x \in \mathcal{G}$ such that $f \leftrightarrow x$ in $\mathbb{Z}_-^d \setminus A$. The intersection of the first two events on the right hand side of (9.102) is independent of the third event, and hence

$$\mathbb{P}_{p,\gamma}^-(W) = \sum_{A:A \ni e} \mathbb{P}_{p,\gamma}^-(C(e) = A, \ e \leftrightarrow \mathcal{G}) \, \mathbb{P}_{p,\gamma}^-(f \leftrightarrow \mathcal{G} \text{ in } \mathbb{Z}_-^d \setminus A). \quad (9.103)$$

Let $M^-(x) = \mathbb{P}^-_{p,\gamma}(x \leftrightarrow \mathcal{G})$, for $x \in \mathbb{Z}^d_-$. Then, by the BK inequality and the fact that the two-point function on $\mathbb{Z}^d_-$ is bounded above by the two-point function on $\mathbb{Z}^d$,

$$\mathbb{P}^-_{p,\gamma}(f \leftrightarrow \mathcal{G} \text{ in } \mathbb{Z}^d_- \setminus A) = M^-(f) - \mathbb{P}^-_{p,\gamma}(f \overset{A}{\leftrightarrow} \mathcal{G})$$
$$\geq M^-(f) - \sum_{y \in A} \tau_{p,0}(f,y) M^-(y). \qquad (9.104)$$

By definition and the BK inequality,

$$M^-(x) = M(p,\gamma) - \mathbb{P}_{p,\gamma}(x \overset{\{0\}}{\longleftrightarrow} \mathcal{G}) \qquad (9.105)$$
$$\geq M(p,\gamma)(1 - \tau_{p,0}(0,x)) \geq M(p,\gamma)(1 - \bar{\mathsf{T}}(p)).$$

In the above, we also used $\tau_{p,0}(0,x) \leq \nabla_p(x) \leq \bar{\mathsf{T}}(p)$, which follows from (9.51) and (9.54).

It follows from (9.103)–(9.105) and the upper bound $M^-(x) \leq M$ that

$$\mathbb{P}^-_{p,\gamma}(W) \geq M(p,\gamma) \qquad (9.106)$$
$$\times \sum_{A:A \ni e} \mathbb{P}^-_{p,\gamma}(C(e) = A, \; e \leftrightarrow \mathcal{G}) \left[1 - \bar{\mathsf{T}}(p) - \sum_{y \in A} \tau_{p,0}(f,y)\right]$$
$$= M(p,\gamma) \left[M^-(e)(1 - \bar{\mathsf{T}}(p)) - \sum_{y \in \mathbb{Z}^d_-} \tau_{p,0}(f,y) \mathbb{P}^-_{p,\gamma}(e \leftrightarrow y, \; e \leftrightarrow \mathcal{G})\right].$$

By the BK inequality,

$$\mathbb{P}^-_{p,\gamma}(e \leftrightarrow y, \; e \leftrightarrow \mathcal{G}) \leq \sum_{w \in \mathbb{Z}^d_-} \tau_{p,0}(e,w)\tau_{p,0}(w,y)M^-(w), \qquad (9.107)$$

and hence, by (9.105)–(9.106),

$$\mathbb{P}^-_{p,\gamma}(W) \geq M(p,\gamma)$$
$$\times \left[M^-(e)(1 - \bar{\mathsf{T}}(p)) - \sum_{y,w \in \mathbb{Z}^d_-} \tau_{p,0}(f,y)\tau_{p,0}(e,w)\tau_{p,0}(w,y)M^-(w)\right]$$
$$\geq M(p,\gamma)^2 \left[(1 - \bar{\mathsf{T}}(p))^2 - \bar{\mathsf{T}}(p)\right]. \qquad (9.108)$$

This completes the proof of (9.98), and hence of (9.99).

Upper bound on X_2. By definition,

$$X_2 = p \sum_{(u,v)} \tau_{p,\gamma}(0,u) \mathbb{E}_{p,\gamma} \left[I[v \Leftrightarrow \mathcal{G}] - \tilde{I}^{\{u,v\}}[v \Leftrightarrow \mathcal{G}]\right]. \qquad (9.109)$$

For the difference of indicators to be nonzero, the double connection from v to $\mathcal{G}$ must be realized via the bond $\{u,v\}$, which therefore must be occupied.

The difference of indicators is therefore bounded above by the indicator that the events $\{v \leftrightarrow \mathcal{G}\}$, $\{u \leftrightarrow \mathcal{G}\}$ and $\{\{u,v\}$ occupied$\}$ occur disjointly. Thus we have

$$\mathbb{E}_{p,\gamma}\left[I[v \Leftrightarrow \mathcal{G}] - \tilde{I}^{\{u,v\}}[v \Leftrightarrow \mathcal{G}] \right] \leq pM(p,\gamma)^2, \qquad (9.110)$$

and hence

$$X_2 \leq p^2 |\Omega| M(p,\gamma)^2 \chi(p,\gamma). \qquad (9.111)$$

But for $v \in \Omega$,

$$p \leq \tau_p(v) \leq \nabla_p(v) \leq \bar{\mathsf{T}}(p) \leq \bar{\mathsf{T}}(p_c), \qquad (9.112)$$

and hence X_2 is negligibly small compared to the lower bound (9.99) on X_1.

Upper bound on X_3. By definition,

$$X_3 = p \sum_{(u,v)} \mathbb{E}_{p,\gamma}\left[\left(\tau_{p,\gamma}(0,u) - \tau_{p,\gamma}^{\tilde{C}^{(u,v)}(v)}(0,u) \right) \tilde{I}^{\{u,v\}}[v \Leftrightarrow \mathcal{G}] \right]. \qquad (9.113)$$

The difference of two-point functions is the expectation of

$$I[0 \leftrightarrow u, 0 \nleftrightarrow \mathcal{G}] - I[0 \leftrightarrow u \text{ in } \mathbb{Z}^d \backslash \tilde{C}^{(u,v)}(v), 0 \nleftrightarrow \mathcal{G}]$$

$$+ I[0 \leftrightarrow u \text{ in } \mathbb{Z}^d \backslash \tilde{C}^{(u,v)}(v), 0 \nleftrightarrow \mathcal{G}] - I[(0 \leftrightarrow u, 0 \nleftrightarrow \mathcal{G}) \text{ in } \mathbb{Z}^d \backslash \tilde{C}^{(u,v)}(v)]$$

$$\leq I[0 \xleftrightarrow{\tilde{C}^{(u,v)}(v)} u, 0 \nleftrightarrow \mathcal{G}], \qquad (9.114)$$

since the second line is non-positive and the first line equals the third line. Since the indicator in (9.113) is bounded above by $I[v \Leftrightarrow \mathcal{G}]$, it follows that

$$X_3 \leq p \sum_{(u,v)} \mathbb{E}_{p,\gamma}\left[\mathbb{P}_{p,\gamma}(0 \xleftrightarrow{\tilde{C}^{(u,v)}(v)} u, 0 \nleftrightarrow \mathcal{G}) \, I[v \Leftrightarrow \mathcal{G}] \right]. \qquad (9.115)$$

To estimate the probability within the expectation, we first note that

$$\mathbb{P}_{p,\gamma}[0 \xleftrightarrow{A} u, 0 \nleftrightarrow \mathcal{G}] \leq \sum_{y \in \mathbb{Z}^d} \mathbb{P}_{p,\gamma}(\{0 \leftrightarrow y\} \circ \{y \leftrightarrow u\}, 0 \nleftrightarrow \mathcal{G}) I[y \in A]. \quad (9.116)$$

By conditioning on $\mathcal{G}$, it can be shown that

$$\mathbb{P}_{p,\gamma}(\{0 \leftrightarrow y\} \circ \{y \leftrightarrow u\}, 0 \nleftrightarrow \mathcal{G}) \leq \tau_{p,\gamma}(0,y)\tau_{p,0}(y,u) \qquad (9.117)$$

(see [104, Lemma 4.3] for further discussion). The important point in (9.117) is that the condition $0 \nleftrightarrow \mathcal{G}$ on the left hand side is retained in the factor $\tau_{p,\gamma}(0,y)$ on the right hand side (but not in $\tau_{p,0}(y,u)$). With (9.115), this gives

$$X_3 \leq p \sum_{(u,v)} \sum_{y \in \mathbb{Z}^d} \tau_{p,\gamma}(0,y)\tau_{p,0}(y,u)\mathbb{E}_{p,\gamma}\left[I[v \Leftrightarrow \mathcal{G}]I[y \in \tilde{C}^{(u,v)}(v)] \right]. \quad (9.118)$$

A further application of BK gives

$$X_3 \leq p \sum_{y \in \mathbb{Z}^d} \tau_{p,\gamma}(0,y) \sum_{(u,v)} \tau_{p,0}(y,u) \sum_{w \in \mathbb{Z}^d} \tau_{p,0}(v,w) \tau_{p,0}(y,w) M(p,\gamma)^2$$

$$\leq p M(p,\gamma)^2 \chi(p,\gamma) |\Omega| \bar{\mathsf{T}}(p), \tag{9.119}$$

using symmetry in the last step. Since $\mathsf{T}(p_c)$ may be assumed to be as small as desired, the upper bound (9.119) is negligibly small compared to the lower bound (9.99) on X_1.

The combination of (9.99), (9.111) and (9.119) completes the proof of (9.87). ∎

10

The Expansion for Percolation

In Sect. 10.1, we derive the expansion for percolation. The expansion produces quantities $\Pi^{(N)}$, and we show in Sect. 10.2 how these quantities can be bounded by Feynman diagrams. Finally, we show in Sect. 10.3 how the diagrams can be bounded in terms of the triangle diagram. The methods of this chapter are due to [94], but we follow the exposition of [32], often verbatim.

We do not prove convergence of the expansion here. Convergence proofs can be found in [32, 91, 94], all based on the bootstrap Lemma 5.9.

10.1 The Expansion

The expansions discussed in Chaps. 3 and 8, for the self-avoiding walk, lattice trees, and lattice animals, each have dual interpretations as arising either from repeated inclusion-exclusion or from expansion of an interaction $\prod_{st}(1 + U_{st})$ followed by resummation involving laces. For percolation, however, only the inclusion-exclusion approach has been used. (For the case of *oriented* percolation, laces can be used; see Sect. 13.2.) In this section, we derive the expansion using repeated inclusion-exclusion. The expansion applies to any graph, finite or infinite, and we will derive the expansion in the setting of an arbitrary connected finite or infinite graph $\mathbb{G}$, which need not even be transitive or regular.[1]

Let $\mathbb{G} = (\mathbb{V}, \mathbb{B})$ be an arbitrary connected graph, finite or infinite, with vertex set $\mathbb{V}$ and edge set $\mathbb{B}$. We consider bond percolation on $\mathbb{G}$. Given $p \in [0, 1]$, we define

$$J(x, y) = pI[\{x, y\} \in \mathbb{B}]$$
$$= p|\Omega|D(x, y) \text{ if } \mathbb{G} \text{ is regular,} \tag{10.1}$$

---

[1] A graph is *regular* if very vertex has the same degree. A *transitive* graph is defined in Sect. 11.4.2.

with $D(x,y)$ equal to the reciprocal of the degree $|\Omega|$ times the indicator that $\{x,y\}$ is an edge. We write $\tau(x,y) = \tau_p(x,y)$ for brevity, and generally drop subscripts indicating dependence on p.

Given a percolation cluster containing 0 and x, we call the connected components that remain after removing all the pivotal bonds for $0 \leftrightarrow x$ *sausages* (vertices are *not* removed when the bonds are removed). See Fig. 10.1. Since they are separated by at least one pivotal bond by definition, no two sausages can have a common vertex. Thus, the sausages are constrained to be mutually avoiding. However, this is a weak constraint, since sausage intersections require a cycle, and cycles are unlikely in high dimensions. The fact that cycles are unlikely also means that sausages tend to be trees. This makes it reasonable to attempt to apply an inclusion-exclusion analysis, where the connection from 0 to x is treated as a random walk path, with correction terms taking into account cycles in sausages and the avoidance constraint between sausages.

The inclusion-exclusion expansion of [94] makes this procedure precise. For each $M = 0, 1, 2, \ldots$, the expansion takes the form

$$\tau(x,y) = \delta_{x,y} + (J*\tau)(x,y) + (\Pi_M * J * \tau)(x,y) + \Pi_M(x,y) + R_M(x,y), \quad (10.2)$$

where the $*$ product denotes matrix multiplication (this reduces to convolution when $\mathbb{G} = \mathbb{Z}^d$ and the graph is translation invariant). The function $\Pi_M : \mathbb{V} \times \mathbb{V} \to \mathbb{R}$ is the central quantity in the expansion, and $R_M(x,y)$ is a remainder term. The dependence of Π_M on M is given by

$$\Pi_M(x,y) = \sum_{N=0}^{M} (-1)^N \Pi^{(N)}(x,y), \quad (10.3)$$

with $\Pi^{(N)}(x,y)$ independent of M. The alternating sign in (10.3) arises via repeated inclusion-exclusion. When the expansion converges, one has

$$\lim_{M \to \infty} \sum_y |R_M(x,y)| = 0. \quad (10.4)$$

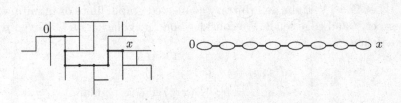

Fig. 10.1. A percolation cluster with a string of 8 sausages joining 0 to x, and a schematic representation of the string of sausages. The 7 pivotal bonds are shown in bold.

This leads to

$$\tau(x,y) = \delta_{x,y} + (J * \tau)(x,y) + (\Pi * J * \tau)(x,y) + \Pi(x,y), \tag{10.5}$$

with $\Pi = \Pi_\infty$ (cf. (8.11) for lattice trees). For a translation invariant bond percolation model on $\mathbb{Z}^d$, taking the Fourier transform of (10.5) and solving for $\hat{\tau}(k)$ gives

$$\hat{\tau}(k) = \frac{1 + \hat{\Pi}(k)}{1 - p|\Omega|\hat{D}(k)[1 + \hat{\Pi}(k)]}. \tag{10.6}$$

This is similar to the formula (3.30) for the self-avoiding walk, but not identical.

The remainder of this section gives the proof of (10.2). Recall Definition 9.13. Also, given a bond configuration, we say that x is *doubly connected to* y, and we write $x \Leftrightarrow y$, if $x = y$ or if there are at least two bond-disjoint paths from x to y consisting of occupied bonds. We denote by $P_{(x,y)}$ the set of directed pivotal bonds for the connection from x to y.

To begin the expansion, we define

$$\Pi^{(0)}(x,y) = \mathbb{P}(x \Leftrightarrow y) - \delta_{x,y} \tag{10.7}$$

and distinguish configurations with $x \leftrightarrow y$ according to whether or not there is a double connection, to obtain

$$\tau(x,y) = \delta_{x,y} + \Pi^{(0)}(x,y) + \mathbb{P}(x \leftrightarrow y, x \not\Leftrightarrow y). \tag{10.8}$$

If x is connected to y, but not doubly, then $P_{(x,y)}$ is nonempty. There is therefore a unique element $(u,v) \in P_{(x,y)}$ (the *first* pivotal bond) such that $x \Leftrightarrow u$, and we can write

$$\mathbb{P}(x \leftrightarrow y, x \not\Leftrightarrow y) = \sum_{(u,v)} \mathbb{P}(x \Leftrightarrow u, \ (u,v) \text{ occupied }, \ (u,v) \in P_{(x,y)}). \tag{10.9}$$

Now comes the essential part of the expansion. Ideally, we would like to factor the probability on the right hand side of (10.9) as

$$\mathbb{P}(x \Leftrightarrow u)\,\mathbb{P}((u,v) \text{ is occupied})\,\mathbb{P}(v \leftrightarrow y) = \left(\delta_{x,u} + \Pi^{(0)}(x,u)\right)J(u,v)\tau(v,y). \tag{10.10}$$

The expression (10.10) leads to (10.2) with $\Pi_M = \Pi^{(0)}$ and $R_M = 0$. However, (10.9) does not factor in this way because the cluster $\tilde{C}^{(u,v)}(u)$ is constrained not to intersect the cluster $\tilde{C}^{(u,v)}(v)$, since (u,v) is pivotal. What we can do is approximate the probability on the right hand side of (10.9) by (10.10), and then attempt to deal with the error term.

For this, we will use the next lemma, which gives an identity for the probability on the right hand side of (10.9). In fact, we will also need a more general identity, involving the following generalizations of the event appearing on the right hand side of (10.9). Let $x, u, v, y \in \mathbb{V}$, and let $A \subset \mathbb{V}$ be nonempty. Then we define the events

$$E'(v, y; A) = \{v \overset{A}{\leftrightarrow} y\} \cap \{\nexists (u', v') \in P_{(v,y)} \text{ such that } v \overset{A}{\leftrightarrow} u'\} \qquad (10.11)$$

and

$$E(x, u, v, y; A) = E'(x, u; A) \cap \{(u, v) \text{ is occupied and pivotal for } x \leftrightarrow y\}. \qquad (10.12)$$

Note that $\{x \leftrightarrow y\} = E'(x, y; \mathbb{V})$, while $E(x, u, v, y; \mathbb{V})$ is the event appearing on the right hand side of (10.9). A version of Lemma 10.1, with $E'(x, u; A)$ replaced by $\{0 \leftrightarrow u\}$ on both sides of (10.13), appeared in (9.78).

Lemma 10.1. *Let $\mathbb{G}$ be a finite or an infinite graph, let $p \in [0, 1]$ be such that there is almost surely no infinite cluster, let $u \in \mathbb{V}$, and let $A \subset \mathbb{V}$ be nonempty. Then*

$$\mathbb{E}\left(I[E(x, u, v, y; A)]\right) = p\mathbb{E}\left(I[E'(x, u; A)] \tau^{\tilde{C}^{\{u,v\}}(x)}(v, y)\right). \qquad (10.13)$$

Proof. The event appearing in the left hand side of (10.13) is depicted in Fig. 10.2. We first observe that the event $E'(x, u; A) \cap \{(u, v) \in P_{(x,y)}\}$ is independent of the occupation status of the bond (u, v). This is true by definition for $\{(u, v) \in P_{(x,y)}\}$, and when (u, v) is pivotal, the occurrence or not of $E'(x, u; A)$ cannot be affected by $\{u, v\}$ since $E'(x, u; A)$ is determined by the occupied paths from x to u and in this case no such path uses the bond $\{u, v\}$. Therefore, the left hand side of the identity in the statement of the lemma is equal to

$$p\mathbb{E}\left(I[E'(x, u; A) \cap \{(u, v) \in P_{(x,y)}\}]\right). \qquad (10.14)$$

By conditioning on $\tilde{C}^{\{u,v\}}(x)$, (10.14) is equal to

$$p \sum_{S:S \ni x} \mathbb{E}\left(I[E'(x, u; A) \cap \{(u, v) \in P_{(x,y)}\} \cap \{\tilde{C}^{\{u,v\}}(x) = S\}]\right), \qquad (10.15)$$

where the sum is over all finite connected sets of vertices S containing x.

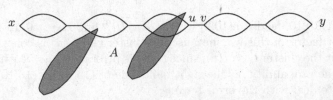

Fig. 10.2. The event $E(x, u, v, y; A)$ of Lemma 10.1. The shaded regions represent the vertices in A. Intersection between A and sausages to the right of (u, v) is permitted.

In (10.15), we can replace

$$\{(u,v) \in P_{(x,y)}\} \cap \{\tilde{C}^{\{u,v\}}(x) = S\} \qquad (10.16)$$

by

$$\{v \leftrightarrow y \text{ in } \mathbb{V}\backslash S\} \cap \{\tilde{C}^{\{u,v\}}(x) = S\}. \qquad (10.17)$$

The event $\{v \leftrightarrow y \text{ in } \mathbb{V}\backslash S\}$ depends only on the occupation status of bonds which do not have an endpoint in S. On the other hand, when $\{v \leftrightarrow y \text{ in } \mathbb{V}\backslash S\} \cap \{\tilde{C}^{\{u,v\}}(x) = S\}$ occurs, the event $E'(x,u;A)$ is determined by the occupation status of bonds which have an endpoint in $S = \tilde{C}^{\{u,v\}}(x)$. Similarly, the event $\{\tilde{C}^{\{u,v\}}(x) = S\}$ depends on bonds which have one or both endpoints in S. Hence, given S, the event $E'(x,u;A) \cap \{\tilde{C}^{\{u,v\}}(x) = S\}$ is independent of the event $\{v \leftrightarrow y \text{ in } \mathbb{V}\backslash S\}$, and therefore (10.15) is equal to

$$p \sum_{S:S \ni x} \mathbb{E}\left(I[E'(x,u;a) \cap \{\tilde{C}^{\{u,v\}}(x) = S\}]\right) \tau_p^S(v,y). \qquad (10.18)$$

Bringing the restricted two-point function inside the expectation, replacing the superscript S by $\tilde{C}^{\{u,v\}}(x)$, and performing the sum over S, gives the desired result. ∎

It follows from (10.9) and Lemma 10.1 that

$$\mathbb{P}(x \leftrightarrow y, \; x \not\Leftrightarrow y) = \sum_{(u,v)} J(u,v)\mathbb{E}\left(I[x \Leftrightarrow u]\, \tau^{\tilde{C}^{(u,v)}(x)}(v,y)\right). \qquad (10.19)$$

On the right hand side, $\tau^{\tilde{C}^{(u,v)}(x)}(v,y)$ is the restricted two-point function *given* the cluster $\tilde{C}^{(u,v)}(x)$ of the expectation $\mathbb{E}$, so that in the expectation defining $\tau^{\tilde{C}^{(u,v)}(x)}(v,y)$, $\tilde{C}^{(u,v)}(x)$ should be regarded as a *fixed* set. The restricted two-point function effectively introduces a second percolation model on a second graph, as in (9.78).

We write

$$\tau^{\tilde{C}^{(u,v)}(x)}(v,y) = \tau(v,y) - \left(\tau(v,y) - \tau^{\tilde{C}^{(u,v)}(x)}(v,y)\right)$$

$$= \tau(v,y) - \mathbb{P}\left(v \xleftarrow{\tilde{C}^{(u,v)}(x)} y\right), \qquad (10.20)$$

insert this into (10.19), and use (10.8) and (10.7) to obtain

$$\tau(x,y) = \delta_{x,y} + \Pi^{(0)}(x,y) + \sum_{(u,v)} \left(\delta_{x,u} + \Pi^{(0)}(x,u)\right) J(u,v)\tau(v,y)$$

$$- \sum_{(u,v)} J(u,v)\mathbb{E}\left(I[x \Leftrightarrow u]\, \mathbb{P}\left(v \xleftarrow{\tilde{C}^{(u,v)}(x)} y\right)\right). \qquad (10.21)$$

With $R_0(x,y)$ equal to the last term on the right hand side of (10.21) (including the minus sign), this proves (10.2) for $M = 0$.

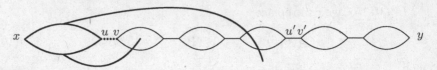

Fig. 10.3. A possible configuration appearing in the second stage of the expansion.

To continue the expansion, we would like to rewrite the final term of (10.21) in terms of a product with the two-point function. A configuration contributing to the expectation in the final term of (10.21) is illustrated schematically in Fig. 10.3, in which the bonds drawn with heavy lines should be regarded as living on a different graph than the bonds drawn with lighter lines. Our goal is to extract a factor $\tau(v', y)$, where v' is shown in Fig. 10.3.

Given a configuration in which $v \xleftrightarrow{A} y$, the *cutting bond* (u', v') is defined to be the first pivotal bond for $v \leftrightarrow y$ such that $v \xleftrightarrow{A} u'$. It is possible that no such bond exists, as for example would be the case in Fig. 10.3 if only the leftmost four sausages were included in the figure, with y in the location currently occupied by u'. Recall the definitions of $E'(v, y; A)$ and $E(x, u, v, y; A)$ in (10.11) and (10.12). By partitioning $\{v \xleftrightarrow{A} y\}$ according to the location of the cutting bond (or the lack of a cutting bond), we obtain the partition

$$\{v \xleftrightarrow{A} y\} = E'(v, y; A) \dot\bigcup \dot\bigcup_{(u',v')} E(v, u', v', y; A), \tag{10.22}$$

which implies that

$$\mathbb{P}(v \xleftrightarrow{A} y) = \mathbb{P}(E'(v, y; A)) + \sum_{(u',v')} \mathbb{P}(E(v, u', v', y; A)). \tag{10.23}$$

Using Lemma 10.1, this gives

$$\mathbb{P}(v \xleftrightarrow{A} y) = \mathbb{P}(E'(v, y; A))$$
$$+ \sum_{(u',v')} J(u', v') \mathbb{E}\left(I[E'(v, u'; A)] \tau^{\tilde{C}^{(u',v')}(v)}(v', y) \right). \tag{10.24}$$

Inserting the identity (10.20) into (10.24), we obtain

$$\mathbb{P}(v \xleftrightarrow{A} y) = \mathbb{P}(E'(v, y; A)) + \sum_{(u',v')} J(u', v') \mathbb{P}(E'(v, u'; A)) \tau(v', y)$$

$$- \sum_{(u',v')} J(u', v') \mathbb{E}_1 \left(I[E'(v, u'; A)] \mathbb{P}_2(v' \xleftarrow{\tilde{C}_1^{(u',v')}(v)} y) \right). \tag{10.25}$$

In the last term on the right hand side, we have introduced subscripts for $\tilde{C}$ and the expectations, to indicate to which expectation $\tilde{C}$ belongs.

Let

$$\Pi^{(1)}(x,y) = \sum_{(u,v)} J(u,v)\, \mathbb{E}_0\Big(I[x \Leftrightarrow u]\mathbb{P}_1\big(E'(v,y;\tilde{C}_0^{(u,v)}(x))\big)\Big). \qquad (10.26)$$

Inserting (10.25) into (10.21), and using (10.26), we have

$$\tau(x,y) = \delta_{x,y} + \Pi^{(0)}(x,y) - \Pi^{(1)}(x,y) \qquad (10.27)$$
$$+ \sum_{(u,v)} \big(\delta_{x,u} + \Pi^{(0)}(x,u) - \Pi^{(1)}(x,u)\big) J(u,v)\, \tau(v,y)$$
$$+ \sum_{(u,v)} J(u,v) \sum_{(u',v')} J(u',v')$$
$$\times \mathbb{E}_0\Big(I[x \Leftrightarrow u]\mathbb{E}_1\big(I[E'(v,u';\tilde{C}_0^{(u,v)}(x))]\mathbb{P}_2(v' \xleftarrow{\tilde{C}_1^{(u',v')}(v)} y)\big)\Big).$$

This proves (10.2) for $M = 1$, with $R_1(x,y)$ given by the last two lines of (10.27).

We now repeat this procedure recursively, rewriting $\mathbb{P}_2(v' \xleftarrow{\tilde{C}_1^{(u',v')}(v)} y)$ using (10.25), and so on. This leads to (10.2), with $\Pi^{(0)}$ and $\Pi^{(1)}$ given by (10.7) and (10.26), and, for $N \geq 2$,

$$\Pi^{(N)}(x,y) = \sum_{(u_0,v_0)} \cdots \sum_{(u_{N-1},v_{N-1})} \Big[\prod_{i=0}^{N-1} J(u_i,v_i)\Big]\mathbb{E}_0 I[x \Leftrightarrow u_0]$$
$$\times \mathbb{E}_1 I[E'(v_0,u_1;\tilde{C}_0)] \cdots \mathbb{E}_{N-1} I[E'(v_{N-2},u_{N-1};\tilde{C}_{N-2})]$$
$$\times \mathbb{E}_N I[E'(v_{N-1},y;\tilde{C}_{N-1})], \qquad (10.28)$$

$$R_M(x,y) = (-1)^{M+1} \sum_{(u_0,v_0)} \cdots \sum_{(u_M,v_M)} \Big[\prod_{i=0}^{M} J(u_i,v_i)\Big]\mathbb{E}_0 I[x \Leftrightarrow u_0]$$
$$\times \mathbb{E}_1 I[E'(v_0,u_1;\tilde{C}_0)] \cdots \mathbb{E}_{M-1} I[E'(v_{M-2},u_{M-1};\tilde{C}_{M-2})]$$
$$\times \mathbb{E}_M\big[I[E'(v_{M-1},u_M;\tilde{C}_{M-1})]\mathbb{P}_{M+1}(v_M \xleftarrow{\tilde{C}_M} y)\big], \qquad (10.29)$$

where we have used the abbreviation $\tilde{C}_j = \tilde{C}_j^{(u_j,v_j)}(v_{j-1})$, with $v_{-1} = x$.

Since

$$\mathbb{P}_{M+1}(v_M \xleftarrow{\tilde{C}_M} y) \leq \tau_p(v_M,y), \qquad (10.30)$$

it follows from (10.28)–(10.29) that

$$|R_M(x,y)| \leq \sum_{u_M,v_M \in \mathbb{V}} \Pi^{(M)}(x,u_M)J(u_M,v_M)\tau_p(v_M,y). \qquad (10.31)$$

This bound is typically used to show that the remainder vanishes in the limit $M \to \infty$, once we can show that $\sum_u \Pi^{(M)}(x, u)$ vanishes in the limit $M \to \infty$.

In deriving the expansion, we have essentially been making an approximation in which $\tau^{\tilde{C}^{\{u,v\}}(0)}(v, x)$ is replaced by $\tau(v, x)$. This is a good approximation when the backbone joining v and x typically does not intersect the cluster $\tilde{C}^{\{u,v\}}(0)$. Above the upper critical dimension, we expect the backbone to behave like Brownian motion, which is 2-dimensional, and the cluster $\tilde{C}^{\{u,v\}}(0)$ to have the dimensionality of an ISE cluster (see Sect. 16.5 below), which is 4-dimensional. These objects generically do not intersect above dimension $6 = 4 + 2$, and this provides an interpretation of the upper critical dimension.

10.2 The Diagrams

In this section, we show how $\Pi^{(N)}$ of (10.28) can be bounded in terms of Feynman diagrams. We follow the presentation of [32], and the method is essentially that of [94, Sect. 2.2]. The results of this section apply to any graph $\mathbb{G} = (\mathbb{V}, \mathbb{B})$, finite or infinite, which need not be transitive nor regular.

Let $\mathbb{P}^{(N)}$ denote the product measure on $N + 1$ copies of percolation on $\mathbb{G}$. By Fubini's Theorem and (10.28),

$$\Pi^{(N)}(x, y) = \sum_{(u_0, v_0)} \cdots \sum_{(u_{N-1}, v_{N-1})} \Big[\prod_{i=0}^{N-1} J(u_i, v_i) \Big] \tag{10.32}$$
$$\times \mathbb{P}^{(N)} \big(\{x \Leftrightarrow u_0\}_0 \cap \big(\bigcap_{i=1}^{N-1} E'(v_{i-1}, u_i; \tilde{C}_{i-1})_i \big) \cap E'(v_{N-1}, y; \tilde{C}_{N-1})_N \big),$$

where, for an event F, we write F_i to denote that F occurs on graph i. To estimate $\Pi^{(N)}(x, y)$ for $N \geq 1$, it is convenient to define the events

$$F_0(x, u_0, w_0, z_1) = \{x \leftrightarrow u_0\} \circ \{x \leftrightarrow w_0\} \circ \{w_0 \leftrightarrow u_0\} \circ \{w_0 \leftrightarrow z_1\}, \tag{10.33}$$

$$F'(v_{i-1}, t_i, z_i, u_i, w_i, z_{i+1}) = \{v_{i-1} \leftrightarrow t_i\} \circ \{t_i \leftrightarrow z_i\} \circ \{t_i \leftrightarrow w_i\} \tag{10.34}$$
$$\circ \{z_i \leftrightarrow u_i\} \circ \{w_i \leftrightarrow u_i\} \circ \{w_i \leftrightarrow z_{i+1}\},$$

$$F''(v_{i-1}, t_i, z_i, u_i, w_i, z_{i+1}) = \{v_{i-1} \leftrightarrow w_i\} \circ \{w_i \leftrightarrow t_i\} \circ \{t_i \leftrightarrow z_i\} \tag{10.35}$$
$$\circ \{t_i \leftrightarrow u_i\} \circ \{z_i \leftrightarrow u_i\} \circ \{w_i \leftrightarrow z_{i+1}\},$$

$$F(v_{i-1}, t_i, z_i, u_i, w_i, z_{i+1}) = F'(v_{i-1}, t_i, z_i, u_i, w_i, z_{i+1}) \tag{10.36}$$
$$\cup F''(v_{i-1}, t_i, z_i, u_i, w_i, z_{i+1}),$$

$$F_N(v_{N-1}, t_N, z_N, y) = \{v_{N-1} \leftrightarrow t_N\} \circ \{t_N \leftrightarrow z_N\} \circ \{t_N \leftrightarrow x\} \circ \{z_N \leftrightarrow y\}.$$
$$(10.37)$$

The events F_0, F', F'', F_N are depicted in Fig. 10.4. Note that

$$F_N(v, t, z, y) = F_0(y, z, t, v). \tag{10.38}$$

By the definition of E' in (10.11),

$$E'(v_{N-1}, y; \tilde{C}_{N-1})_N \subset \bigcup_{z_N \in \tilde{C}_{N-1}} \bigcup_{t_N \in \mathbb{V}} F_N(v_{N-1}, t_N, z_N, y)_N. \tag{10.39}$$

Indeed, viewing the connection from v_{N-1} to y as a string of sausages beginning at v_{N-1} and ending at y, for the event E' to occur there must be a vertex $z_N \in \tilde{C}_{N-1}$ that lies on the last sausage, on a path from v_{N-1} to y. (In fact, both "sides" of the sausage must contain a vertex in $\tilde{C}_{N-1}$, but we do not need or use this.) This leads to (10.39), with t_N representing the other endpoint of the sausage that terminates at y.

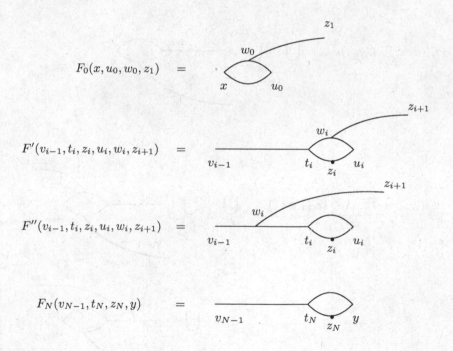

Fig. 10.4. Diagrammatic representations of the events $F_0(x, u_0, w_0, z_1)$, $F'(v_{i-1}, t_i, z_i, u_i, w_i, z_{i+1})$, $F''(v_{i-1}, t_i, z_i, u_i, w_i, z_{i+1})$, $F_N(v_{N-1}, t_N, z_N, y)$. Lines indicate disjoint connections.

Assume, for the moment, that $N \geq 2$. The condition in (10.39) that $z_N \in \tilde{C}_{N-1}$ is a condition on the graph $N - 1$ that must be satisfied in conjunction with the event $E'(v_{N-2}, u_{N-1}; \tilde{C}_{N-2})_{N-1}$. It is not difficult to see that for $i \in \{1, \ldots, N-1\}$,

$$E'(v_{i-1}, u_i; \tilde{C}_{i-1})_i \cap \{z_{i+1} \in \tilde{C}_i\} \subset \bigcup_{z_i \in \tilde{C}_{i-1}} \bigcup_{t_i, w_i \in \mathbb{V}} F(v_{i-1}, t_i, z_i, u_i, w_i, z_{i+1})_i.$$

(10.40)

See Fig. 10.5 for a depiction of the inclusions in (10.39) and (10.40). Informally, for (10.40), any path from v_{i-1} to u_i must contain a vertex with a disjoint connection to z_{i+1}, and this can occur in the two topologically distinct manners illustrated in Fig. 10.5. The definition of E' requires the existence of a vertex $z_i \in \tilde{C}_{i-1}$ on *every* path from t_i to u_i, where the latter are the endpoints of the last sausage, and so, in particular, such a vertex can be found on the "other" side of the sausage, in the case where z_{i+1} connects to the last sausage.

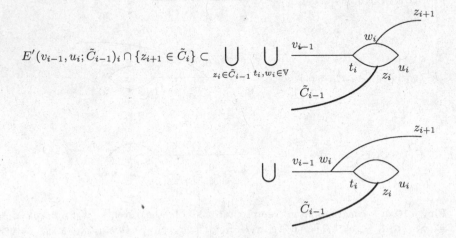

Fig. 10.5. Diagrammatic representations of the inclusions in (10.39) and (10.40).

With an appropriate treatment for graph 0, (10.39) and (10.40) lead to

$$\{x \Leftrightarrow u_0\}_0 \cap \Big(\bigcap_{i=1}^{N-1} E'(v_{i-1}, u_i; \tilde{C}_{i-1})_i \Big) \cap E'(v_{N-1}, y; \tilde{C}_{N-1})_N \qquad (10.41)$$

$$\subset \bigcup_{\vec{t}, \vec{w}, \vec{z}} \Big(F_0(x, u_0, w_0, z_1)_0 \cap \Big(\bigcap_{i=1}^{N-1} F(v_{i-1}, t_i, z_i, u_i, w_i, z_{i+1})_i \Big)$$

$$\cap F_N(v_{N-1}, t_N, z_N, y)_N \Big), \qquad (10.42)$$

where $\vec{t} = (t_1, \ldots, t_N)$, $\vec{w} = (w_0, \ldots, w_{N-1})$ and $\vec{z} = (z_1, \ldots, z_N)$. Therefore,

$$\Pi^{(N)}(x, y) \leq \sum \left[\prod_{i=0}^{N-1} J(u_i, v_i) \right] \mathbb{P}_p(F_0(x, u_0, w_0, z_1))$$

$$\times \prod_{i=1}^{N-1} \mathbb{P}_p(F(v_{i-1}, t_i, z_i, u_i, w_i, z_{i+1})) \mathbb{P}_p(F_N(v_{N-1}, t_N, z_N, y)),$$

$$(10.43)$$

where the summation is over all vertices $z_1, \ldots, z_N$, $t_1, \ldots, t_N$, $w_0, \ldots, w_{N-1}$, $u_0, \ldots, u_{N-1}$, $v_0, \ldots, v_{N-1}$. The probability in (10.43) factors because the events $F_0, \ldots, F_N$ are events on different percolation models. Each probability in (10.43) can be estimated using the BK inequality. The result is that each of the connections $\{a \leftrightarrow b\}$ present in the events F_0, F and F_N is replaced by a two-point function $\tau_p(a, b)$. This results in a large sum of two-point functions.

To organize a large sum of this form, we let

$$\tilde{\tau}_p(x, y) = (J * \tau_p)(x, y)$$

$$= p|\Omega|(D * \tau_p)(x, y) \text{ if } \mathbb{G} \text{ is regular}, \qquad (10.44)$$

and define

$$A_3(s, u, v) = \tau_p(s, v)\tau_p(s, u)\tau_p(u, v), \qquad (10.45)$$

$$B_1(s, t, u, v) = \tilde{\tau}_p(t, v)\tau_p(s, u), \qquad (10.46)$$

$$B_2(u, v, s, t) = \tau_p(u, v)\tau_p(u, t)\tau_p(v, s)\tau_p(s, t)$$

$$+ \delta_{v,s}\tau_p(u, t) \sum_{a \in \mathbb{V}} \tau_p(s, a)\tau_p(a, u)\tau_p(a, t). \qquad (10.47)$$

The two terms in B_2 arise from the two events F' and F'' in (10.36). We will write them as $B_2^{(1)}$ and $B_2^{(2)}$, respectively. The above quantities are represented diagrammatically in Fig. 10.6. In the diagrams, a line joining a and b represents $\tau_p(a, b)$. In addition, small bars are used to distinguish a line that represents $\tilde{\tau}_p$, in B_1.

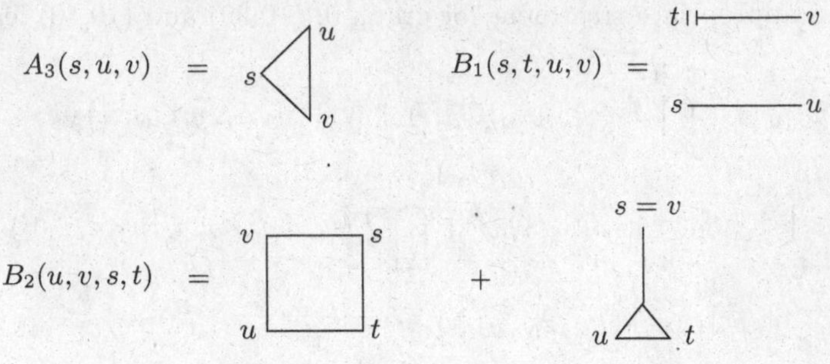

$$A_3(s, u, v) \quad = \qquad\qquad\qquad B_1(s, t, u, v) \quad =$$

$$B_2(u, v, s, t) \quad = \qquad\qquad\qquad +$$

Fig. 10.6. Diagrammatic representations of $A_3(s, u, v)$, $B_1(s, t, u, v)$ and $B_2(u, v, s, t)$.

Application of the BK inequality yields

$$\mathbb{P}_p(F_0(x, u_0, w_0, z_1)) \leq A_3(x, u_0, w_0)\tau_p(w_0, z_1), \qquad (10.48)$$

$$\sum_{v_{N-1}} J(u_{N-1}, v_{N-1})\mathbb{P}_p(F_N(v_{N-1}, t_N, z_N, y)) \qquad (10.49)$$

$$\leq \frac{B_1(w_{N-1}, u_{N-1}, z_N, t_N)}{\tau_p(w_{N-1}, z_N)} A_3(y, t_N, z_N),$$

$$\sum_{v_{i-1}} J(u_{i-1}, v_{i-1})\mathbb{P}_p(F'(v_{i-1}, t_i, z_i, u_i, w_i, z_{i+1}))$$

$$\leq \frac{B_1(w_{i-1}, u_{i-1}, z_i, t_i)}{\tau_p(w_{i-1}, z_i)} B_2^{(1)}(z_i, t_i, w_i, u_i)\tau_p(w_i, z_{i+1}), \qquad (10.50)$$

$$\sum_{v_{i-1}, t_i} J(u_{i-1}, v_{i-1})\mathbb{P}_p(F''(v_{i-1}, t_i, z_i, u_i, w_i, z_{i+1}))$$

$$\leq \frac{B_1(w_{i-1}, u_{i-1}, z_i, w_i)}{\tau_p(w_{i-1}, z_i)} B_2^{(2)}(z_i, w_i, w_i, u_i)\tau_p(w_i, z_{i+1}). \qquad (10.51)$$

Since the second and the third arguments of $B_2^{(2)}$ are equal by virtue of the Kronecker delta in (10.47), we can combine (10.50)–(10.51) to obtain

$$\sum_{v_{i-1}, t_i} J(u_{i-1}, v_{i-1})\mathbb{P}_p(F(v_{i-1}, t_i, z_i, u_i, w_i, z_{i+1}))$$

$$\leq \sum_{t_i} \frac{B_1(w_{i-1}, u_{i-1}, z_i, t_i)}{\tau_p(w_{i-1}, z_i)} B_2(z_i, t_i, w_i, u_i)\tau_p(w_i, z_{i+1}). \qquad (10.52)$$

Upon substitution of the bounds on the probabilities in (10.48), (10.49) and (10.52) into (10.43), the ratios of two-point functions form a telescoping product that disappears. After relabelling the summation indices, (10.43) becomes

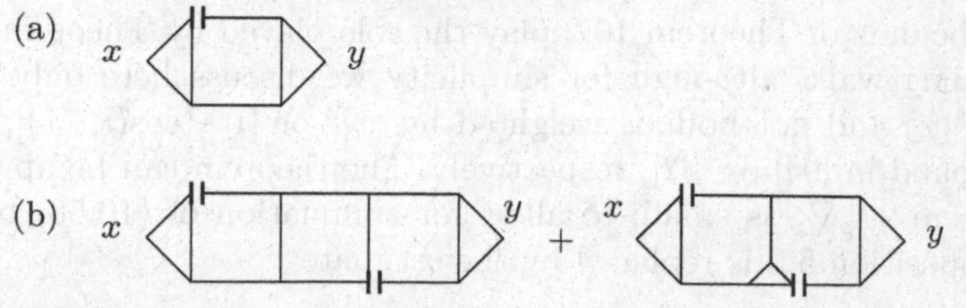

(a)

(b)

Fig. 10.7. The diagrams bounding (a) $\Pi^{(1)}(x,y)$ and (b) $\Pi^{(2)}(x,y)$.

$$\Pi^{(N)}(x,y) \leq \sum_{\vec{s},\vec{t},\vec{u},\vec{v}} A_3(x,s_1,t_1) \prod_{i=1}^{N-1} \left[B_1(s_i,t_i,u_i,v_i) B_2(u_i,v_i,s_{i+1},t_{i+1}) \right]$$

$$\times B_1(s_N,t_N,u_N,v_N) A_3(u_N,v_N,y). \tag{10.53}$$

The bound (10.53) is valid for $N \geq 1$, and the summation is over all $s_1,\ldots,s_N$, $t_1,\ldots,t_N$, $u_1,\ldots,u_N$, $v_1,\ldots,v_N$. For $N = 1, 2$, the right hand side is represented diagrammatically in Fig. 10.7. In the diagrams, unlabelled vertices are summed over $\mathbb{V}$.

10.3 Diagrammatic Estimates for Percolation

In this section, we indicate how the Feynman diagrams can be bounded in terms of quantities related to the triangle diagram. We now specialize to the case where the graph $\mathbb{G}$ has vertex set $\mathbb{Z}^d$ and edge set given by $y - x \in \Omega$ with either (1.1) or (1.2), and make use of the additive structure and the $x \mapsto -x$ symmetry. We will write $\tau_p(y - x)$ in place of $\tau_p(x,y)$, $p|\Omega|D(y-x)$ in place of $J(x,y)$, and $\Pi^{(N)}(y-x)$ in place of $\Pi^{(N)}(x,y)$. We recall the definitions of $\nabla_p(x)$ and $\tilde{\tau}_p(x)$ in (9.51) and (10.44), and define

$$T_p(x) = (\tau_p * \tau_p * \tilde{\tau}_p)(x), \tag{10.54}$$

$$T_p = \sup_{x \in \mathbb{Z}^d} T_p(x), \tag{10.55}$$

$$\bar{\nabla}_p = \sup_{x \in \mathbb{Z}^d} \nabla_p(x) = \sup_{x \in \mathbb{Z}^d} (\tau_p * \tau_p * \tau_p)(x). \tag{10.56}$$

Theorem 10.2. *For $N = 0$,*

$$\sum_{x \in \mathbb{Z}^d} \Pi^{(0)}(x) \leq T_p. \tag{10.57}$$

For $N \geq 1$,

$$\sum_{x \in \mathbb{Z}^d} \Pi^{(N)}(x) \leq \bar{\nabla}_p (2 T_p \bar{\nabla}_p)^N. \tag{10.58}$$

The bounds of Theorem 10.2 play the role played by Theorem 4.1 for the self-avoiding walk, although for simplicity we discuss here only bounds on $\sum_x \Pi^{(N)}(x)$ and not bounds weighted by $|x|^2$ or $[1 - \cos(k \cdot x)]$. The latter can be found in [94] or [32], respectively. The theorem can be applied in settings where $2T_p \bar{\nabla}_p$ is small, to allow for summation of (10.58) over N. For this, Proposition 5.3 is replaced by the estimate

$$
\int_{[-\pi,\pi]^d} \frac{\hat{D}(k)^2}{[1 - \hat{D}(k)]^3} \frac{\mathrm{d}^d k}{(2\pi)^d} \leq \beta,
\tag{10.59}
$$

with $\beta = K(d-6)^{-1}$ for the nearest-neighbour model, and $\beta = KL^{-d}$ for the spread-out model in dimensions $d > 6$ (see [32] for this perspective).

Proof of Theorem 10.2. We first prove (10.57). By (10.7) and the BK inequality,

$$
\Pi^{(0)}(x) = \mathbb{P}(0 \Leftrightarrow x) - \delta_{0,x} \leq \tau_p(x)^2 - \delta_{0,x}.
\tag{10.60}
$$

For $x \neq 0$, the event $\{0 \leftrightarrow x\}$ is the union over neighbors y of the origin of $\{\{0,y\} \text{ occupied}\} \circ \{y \leftrightarrow x\}$. Thus, by the BK inequality,

$$
\tau_p(x) \leq p|\Omega|(D * \tau_p)(x) = \tilde{\tau}_p(x) \quad (x \neq 0).
\tag{10.61}
$$

Therefore,

$$
\sum_{x \in \mathbb{Z}^d} \Pi^{(0)}(x) \leq \sum_{x \in \mathbb{Z}^d} \tau_p(x) \tilde{\tau}_p(x) \leq T_p(0).
\tag{10.62}
$$

This proves (10.57).

Next, we prove (10.58). For $N \geq 1$, let

$$
\Psi^{(N)}(s_{N+1}, t_{N+1})
\tag{10.63}
$$

$$
= \sum_{\vec{s},\vec{t},\vec{u},\vec{v}} A_3(0, s_1, t_1) \prod_{i=1}^{N} \left[B_1(s_i, t_i, u_i, v_i) B_2(u_i, v_i, s_{i+1}, t_{i+1}) \right],
$$

where the sum is over $s_1, \ldots, s_N, t_1, \ldots, t_N, u_1, \ldots, u_N, v_1, \ldots, v_N$. For convenience, we define $\Psi^{(0)}(x,y) = A_3(0,x,y)$, so that for $N \geq 1$,

$$
\Psi^{(N)}(x,y) = \sum_{u_N, v_N, s_N, t_N} \Psi^{(N-1)}(s_N, t_N) B_1(s_N, t_N, u_N, v_N) B_2(u_N, v_N, x, y).
\tag{10.64}
$$

Since

$$
\sum_x A_3(u_N, v_N, x) \leq \sum_{x,y} B_2(u_N, v_N, x, y),
\tag{10.65}
$$

it follows from (10.53) that

$$
\sum_x \Pi^{(N)}(x) \leq \sum_{x,y} \Psi^{(N)}(x,y),
\tag{10.66}
$$

and bounds on $\Pi^{(N)}$ can be obtained from bounds on $\Psi^{(N)}$. We prove bounds on $\Psi^{(N)}$, and hence on $\Pi^{(N)}$, by induction on N.

The induction hypothesis is that

$$\sum_{x,y} \Psi^{(N)}(x,y) \leq \bar{\nabla}_p (2T_p \bar{\nabla}_p)^N. \tag{10.67}$$

For $N = 0$, (10.67) is true since

$$\sum_{x,y} A_3(0,x,y) \leq \bar{\nabla}_p. \tag{10.68}$$

If we assume (10.67) is valid for $N - 1$, then by (10.64),

$$\sum_{x,y} \Psi^{(N)}(x,y) \leq \Big(\sum_{s_N,t_N} \Psi^{(N-1)}(s_N,t_N) \Big) \tag{10.69}$$

$$\times \Big(\sup_{s_N,t_N} \sum_{u_N,v_N,x,y} B_1(s_N,t_N,u_N,v_N) B_2(u_N,v_N,x,y) \Big),$$

and (10.67) then follows once we prove that

$$\sup_{s,t} \sum_{u,v,x,y} B_1(s,t,u,v) B_2(u,v,x,y) \leq 2T_p \bar{\nabla}_p. \tag{10.70}$$

It remains to prove (10.70). There are two terms, due to the two terms in (10.47), and we bound each term separately. The first term is bounded as

$$\sup_{s,t} \sum_{u,v,x,y} \tilde{\tau}_p(v-t)\tau_p(u-s)\tau_p(y-u)\tau_p(x-v)\tau_p(v-u)\tau_p(x-y)$$

$$= \sup_{s,t} \sum_{u,v} \tilde{\tau}_p(v-t)\tau_p(u-s)\tau_p(v-u)\Big(\sum_{x,y} \tau_p(y-u)\tau_p(x-v)\tau_p(x-y) \Big)$$

$$\leq \bar{\nabla}_p \sup_{s,t} \sum_{u,v} \tilde{\tau}_p(v-t)\tau_p(u-s)\tau_p(v-u)$$

$$= T_p \bar{\nabla}_p. \tag{10.71}$$

The second term is bounded similarly, making use of symmetry, by

$$\sup_{s,t} \sum_{u,v,x,y,a} \tilde{\tau}_p(v-t)\tau_p(u-s)\delta_{v,x}\tau_p(y-u)\tau_p(x-a)\tau_p(u-a)\tau_p(y-a)$$

$$= \sup_{s,t} \sum_{a,y,u} \big((\tilde{\tau}_p * \tau)(a-t)\tau_p(u-s) \big) \big(\tau_p(y-u)\tau_p(u-a)\tau_p(y-a) \big)$$

$$= \sup_{s,t} \sum_{y',a'} T_p(a'+s-t)\tau_p(y')\tau_p(a')\tau_p(y'-a')$$

$$\leq \big(\sup_{a',s,t} T_p(a'+s-t) \big) \big(\sum_{y',a'} \tau_p(y')\tau_p(a')\tau_p(y'-a') \big)$$

$$\leq T_p \bar{\nabla}_p, \tag{10.72}$$

where $a' = a - u$, $y' = y - u$. This completes the proof of (10.70) and hence of (10.58). ∎

11

Results for Percolation

In this chapter, we survey results that have been obtained for percolation using the expansion of Chap. 10, and extensions of this expansion. For the results on $\mathbb{Z}^d$ with bonds $\{x, y\}$ specified by $y - x \in \Omega$, we consider the nearest-neighbour model with Ω given by (1.1) and the spread-out model with Ω given by (1.2).

11.1 Critical Exponents

The following infrared bound was proved in [94] (abbreviated versions appeared in [92, 93]).

Theorem 11.1. *There is a constant independent of p and k such that*

$$\hat{\tau}_p(k) \leq \text{const} \, |k|^{-2} \tag{11.1}$$

uniformly in $p < p_c$ and $k \in [-\pi, \pi]^d$, for the nearest-neighbour model in dimensions $d \geq d_0$ for some $d_0 \geq 6$, and for the spread-out model if $d > 6$ and $L \geq L_0$ for some $L_0 = L_0(d) \gg 1$.

For the spread-out model, much more general models are actually considered in [94]. In particular, models with bonds of unbounded length are allowed if the occupation probability decays sufficiently rapidly with bond length. For the nearest-neighbour model, the constant d_0 in Theorem 11.1 was found to be $d_0 = 48$ in [94], and this was improved to $d_0 = 19$ in [100], although the lengthy numerical calculations that produce these particular numbers were never published.

The triangle condition follows immediately from the infrared bound. To see this, we first note from (9.51), (9.53) and the monotone convergence theorem[1] that the infrared bound gives

---

[1] Continuity of $\tau_p(x)$ for all p is established in [11].

$$\mathsf{T}(p_c) = \lim_{p \to p_c^-} \sum_{x,y \in \mathbb{Z}^d} \tau_p(x)\tau_p(y-x)\tau_p(y)$$

$$= \lim_{p \to p_c^-} \int_{[-\pi,\pi]^d} \hat{\tau}_p(k)^3 \frac{d^d k}{(2\pi)^d}$$

$$\leq \text{const} \int_{[-\pi,\pi]^d} \frac{1}{|k|^6} \frac{d^d k}{(2\pi)^d}, \tag{11.2}$$

and the integral is finite for $d > 6$. In addition, it is shown in [94] that $\bar{\mathsf{T}}(p_c) \leq O(d^{-1})$ for the nearest-neighbour model (see [94, Lemma 4.4]), and the results of [94] can be extended in a straightforward way to prove that $\bar{\mathsf{T}}(p_c) \leq O(L^{-d})$ for the spread-out model (see [118, Lemma 3.1]). This is important for application of Theorem 10.2.

It follows from the differential inequalities of Sects. 9.2–9.3 (see Corollaries 9.3 and 9.8 and Theorems 9.5 and 9.10) that under the hypotheses of Theorem 11.1,

$$\chi(p) \simeq (p_c - p)^{-1} \qquad \text{as } p \to p_c^-, \tag{11.3}$$

$$M(p_c, \gamma) \simeq \gamma^{1/2} \qquad \text{as } \gamma \to 0^+, \tag{11.4}$$

$$\theta(p) \simeq (p - p_c) \qquad \text{as } p \to p_c^+. \tag{11.5}$$

(As defined in (2.32), the symbol $\simeq$ implies upper and lower bounds with possibly different constants.) This shows that $\gamma = \beta = 1$ and $\delta = 2$. Also, it was shown in [165] that the triangle condition implies that

$$\frac{\mathbb{E}_p(|C(0)|^{m+1})}{\mathbb{E}_p(|C(0)|^m)} \simeq (p_c - p)^{-2} \quad \text{as } p \to p_c^-, \text{ for } m = 1, 2, \ldots, \tag{11.6}$$

so the *gap exponent* Δ exists and takes the value $\Delta = 2$ under the hypotheses of Theorem 11.1.

Using a novel argument, it was shown in [89] that

$$\xi(p) \simeq (p_c - p)^{-1/2} \quad \text{as } p \to p_c^- \tag{11.7}$$

under the hypotheses of Theorem 11.1, which is to say that the critical exponent ν for the correlation length (9.4) exists and equals $\frac{1}{2}$.

Concerning the critical exponent α for the free energy, less is known. In [210], it is shown that for nearest-neighbour bond percolation on $\mathbb{Z}^d$, given any non-negative integer n, there exists a dimension d_n such that $\kappa(p)$ is n times continuously differentiable for $p \in [0, p_c]$, and hence the derivative $\kappa^{(n)}(p)$ is uniformly bounded on $[0, p_c]$ if $d \geq d_n$. The derivatives at p_c are interpreted as left derivatives. The possibility is not ruled out in [210] that $\kappa^{(n)}(p_c^-)$ is equal to zero, although this is not expected to be the case. Comparing with (9.14), the conclusion of [210] is that $\alpha \leq -1$ in high dimensions, consistent with $\alpha = -1$. The methods of [210] apply also to the spread-out model, but the relevance of this is diminished by the fact that the dimension

d_n grows with n, and results for all $d > 6$ have not been obtained. For a regular infinite tree of degree $|\Omega| \geq 3$, it is known (see [210]) that the analogue $\kappa_{\text{tree}}(p)$ of $\kappa(p)$ is twice differentiable on $[0, 1]$, $\kappa_{\text{tree}}'''(p)$ has a jump discontinuity at $p = \frac{1}{|\Omega|-1}$, and $\kappa_{\text{tree}}^{(n)}(p)$ is bounded uniformly in $p \neq \frac{1}{|\Omega|-1}$ for each n. The exponent α is generally tricky, and remains mysterious even for $d = 2$, where other exponents are understood [191].

Concerning the critical exponent η, recalling the characterization of η as $\hat{\tau}_{p_c}(k) \sim c|k|^{\eta-2}$, the infrared bound of Theorem 11.1 can be interpreted as a k-space statement of the mean-field bound $\eta \geq 0$. This has been greatly extended, in two ways.

Firstly, the following theorem from [91] gives an x-space statement that $\eta = 0$. Note that it is, in general, not possible to conclude asymptotic behaviour $|x|^{2-d}$ directly from asymptotic behaviour $|k|^{-2}$ of the Fourier transform, without additional regularity of the Fourier transform. A detailed discussion of this point can be found in [158, pp. 32–33]. Thus, additional input beyond Theorem 11.1 is needed to prove the following theorem. The proof of Theorem 11.2 actually proceeds directly in x-space, without using Fourier transforms, and is not a corollary of Theorem 11.1.

Theorem 11.2. *(a) [90] For $d \geq 19$, there is a constant a' depending on d, such that the critical two-point function of nearest-neighbour percolation obeys, as $|x| \to \infty$,*

$$\tau_{p_c}(x) = \frac{a'}{|x|^{d-2}} \left[1 + O\left(\frac{1}{|x|^{2/d}}\right) \right]. \tag{11.8}$$

(b) [91] Fix any $\alpha > 0$ (think of α small). For $d > 6$, there is a finite constant a depending on d and L, and an L_0 depending on d and α, such that for $L \geq L_0$ the spread-out model of strictly self-avoiding walk obeys, as $|x| \to \infty$,

$$\tau_{p_c}(x) = \frac{a}{|x|^{d-2}} \left[1 + O\left(\frac{1}{|x|^{[2\wedge2(d-6)]-\alpha}}\right) \right]. \tag{11.9}$$

The constant in the error term is uniform in x but may depend on L and α; a more careful statement in this regard is given in [91].

Secondly, in k-space, the infrared bound has been extended to a *joint* asymptotic formula as $k \to 0$ and $\gamma \to 0^+$, at $p = p_c$, involving both $\eta = 0$ and $\delta = 2$. We use translation invariance to rewrite (9.89) as

$$\tau_{p,\gamma}(x) = \mathbb{P}_{p,\gamma}(0 \leftrightarrow x, 0 \nleftrightarrow \mathcal{G}), \tag{11.10}$$

and consider the Fourier transform

$$\hat{\tau}_{p,\gamma}(k) = \sum_{x \in \mathbb{Z}^d} \tau_{p,\gamma}(x) e^{ik \cdot x}. \tag{11.11}$$

Theorem 11.3. *Let $p = p_c$, $k \in [-\pi, \pi]^d$, $\gamma \in (0, 1]$. For nearest-neighbour bond percolation with d sufficiently large, and for spread-out bond percolation with $d > 6$ and L sufficiently large, there are functions $\epsilon_1(z)$ and $\epsilon_2(k)$ with $\lim_{\gamma \to 0^+} \epsilon_1(\gamma) = \lim_{k \to 0} \epsilon_2(k) = 0$, and constants C_2 and D_2 depending on d and L, such that*

$$\hat{\tau}^{(2)}_{p_c,\gamma}(k) = \frac{C_2}{D_2^2|k|^2 + 2^{3/2}\gamma^{1/2}} \left[1 + \epsilon(\gamma, k)\right] \tag{11.12}$$

with $|\epsilon(\gamma, k)| \leq \epsilon_1(\gamma) + \epsilon_2(k)$. In addition, the limit $\hat{\tau}_{p_c,0}(k) = \lim_{\gamma \to 0^+} \hat{\tau}_{p_c,\gamma}(k)$ exists and is finite for $k \neq 0$, and obeys

$$\hat{\tau}_{p_c,0}(k) = \frac{C_2}{D_2^2|k|^2} \left[1 + \epsilon_2(k)\right]. \tag{11.13}$$

The factor $2^{3/2}$ in (11.12) is present to agree with our convention in Chap. 16, where (11.12) is related to integrated super-Brownian excursion. By an appeal to universality, Theorem 11.3 strongly suggests that (11.12) and (11.13) should also be valid for the nearest-neighbour model for all $d > 6$.

Equation (11.13) is a k-space statement that $\eta = 0$. There is good reason to be careful with the limit $\gamma \to 0^+$ in (11.13). In fact, according to Theorem 11.2, $\tau_{p_c,0}(x)$ decays like $|x|^{2-d}$. Therefore it is not summable in x, and its Fourier transform is not well-defined without some interpretation. We use the interpretation $\hat{\tau}_{p_c,0}(k) = \lim_{\gamma \to 0^+} \hat{\tau}_{p_c,\gamma}(k)$ because $\tau_{p_c,0}(x)$ is then the inverse Fourier transform of $\hat{\tau}_{p_c,0}(k)$. In fact, using monotone convergence in the first step, and using (11.12) and the dominated convergence theorem in the last step, we have

$$\tau_{p_c,0}(x) = \lim_{\gamma \to 0^+} \tau_{p_c,\gamma}(x)$$

$$= \lim_{\gamma \to 0^+} \int_{[-\pi,\pi]^d} \hat{\tau}_{p_c,\gamma}(k) e^{-ik \cdot x} \frac{d^d k}{(2\pi)^d}$$

$$= \int_{[-\pi,\pi]^d} \hat{\tau}_{p_c,0}(k) e^{-ik \cdot x} \frac{d^d k}{(2\pi)^d}. \tag{11.14}$$

Recalling the definition of $\chi(p, \gamma)$ in (9.29), and the representation (9.90), we have $\hat{\tau}_{p_c,\gamma}(0) = \chi(p_c, \gamma)$. Setting $k = 0$ in (11.12) gives

$$\chi(p_c, \gamma) = \hat{\tau}_{p_c,\gamma}(0) = \gamma^{-1/2} \left[2^{-3/2} C_2 + o(1)\right], \tag{11.15}$$

as $\gamma \to 0^+$. Integration of this with respect to γ, using $M(p_c, 0) = 0$ and again recalling (9.29), gives

$$M(p_c, \gamma) = \int_0^\gamma \frac{1}{1-t} \chi(p_c, t) dt = \gamma^{1/2} \left[2^{-1/2} C_2 + o(1)\right]. \tag{11.16}$$

This improves (11.4) to a a statement that $\delta = 2$ in the sense of an asymptotic formula.

In [105], a stronger statement that the critical exponent δ equals 2 is proved for the nearest-neighbour model in sufficiently high dimensions. Namely, it is shown that

$$\mathbb{P}_{p_c}(|C(0)| = n) = \frac{C_2}{\sqrt{8\pi}} \frac{1}{n^{3/2}}[1 + O(n^{-\epsilon})], \tag{11.17}$$

for any fixed $\epsilon < \frac{1}{2}$, if d is large enough. In fact, much more is shown in [105], and the results are summarized below in Sect. 16.5.

The proofs of Theorem 11.3 and (11.17) are based on an extension of the expansion of Chap. 10 from an expansion for $\tau_p(x) = \tau_{p,0}(x)$ to an expansion for $\tau_{p,\gamma}(x)$. This new expansion relies heavily on the probabilistic interpretation (9.89) of $\tau_{p,\gamma}(x)$, and is not a minor extension of the expansion of Chap. 10. Details can be found in [104, 105].

Finally, we mention that it is indicated in [198] that the triangle condition can been extended to continuum percolation [160], if the dimension d is large enough, yielding existence of some critical exponents.

11.2 The Critical Value

Several authors have considered the asymptotic behaviour of the critical value $p_c(d)$ of bond percolation on $\mathbb{Z}^d$, in the limit $d \to \infty$. Bollobás and Kohayakawa [27], Gordon [82], Kesten [143] and Hara and Slade [94] proved that $p_c(d)$ is equal to $\frac{1}{2d}$ plus an error term of size $O((\log d)^2 d^{-2})$, $O(d^{-65/64})$, $O((\log \log d)^2(d \log d)^{-1})$ and $O(d^{-2})$, respectively. Recently, Alon, Benjamini and Stacey [17] gave an alternate and relatively short proof that $p_c(d)$ is asymptotic to $\frac{1}{2d}$ as $d \to \infty$, with an estimate $o(d^{-1})$ for the error. The expansion

$$p_c(d) = \frac{1}{2d} + \frac{1}{(2d)^2} + \frac{7}{2(2d)^3} + \frac{16}{(2d)^4} + \frac{103}{(2d)^5} + \cdots \tag{11.18}$$

was reported in [78], but with no rigorous bound on the remainder.

The lace expansion was used in [101, 102] to prove that

$$p_c(d) = \frac{1}{2d} + \frac{1}{(2d)^2} + \frac{7}{2(2d)^3} + O((2d)^{-4}), \tag{11.19}$$

and a simpler proof of this fact, again using the lace expansion, was given in [123]. Moreover, it was proved in [122] that there exist rational numbers a_i such that $p_c(d)$ has an asymptotic expansion

$$p_c(d) \sim \sum_{n=1}^{\infty} \frac{a_i}{(2d)^i}. \tag{11.20}$$

Presumably, this expansion is divergent (though there is no proof of this), and the meaning of (11.20) is that for each $M \geq 1$,

$$p_c(d) = \sum_{i=1}^{M} \frac{a_i}{(2d)^i} + O\left(\frac{1}{(2d)^{M+1}}\right). \tag{11.21}$$

The constant in the error term is permitted to depend on M, so (11.21) does not imply convergence.

The basic method of the proofs of (11.19) and (11.20) is similar to that employed to obtain the asymptotic expansion for the connective constant in (2.8), which was discussed around (5.66) and in Exercise 5.18. The starting point is the formula

$$\chi(p) = \frac{1 + \hat{\Pi}_p(0)}{1 - 2dp[1 + \hat{\Pi}_p(0)]} \tag{11.22}$$

for the susceptibility, which follows from (10.6) and $\chi(p) = \hat{\tau}_p(0)$. The function $\hat{\Pi}_p(0)$ is well-behaved up to and including $p = p_c$, and the critical point is characterized by the equation

$$1 - 2dp_c[1 + \hat{\Pi}_{p_c}(0)] = 0, \tag{11.23}$$

or, equivalently,

$$p_c = \frac{1}{2d}[1 + \hat{\Pi}_{p_c}(0)]^{-1}. \tag{11.24}$$

This equation can then be studied in a recursive fashion, using estimates on $\hat{\Pi}_{p_c}(0)$.

For the spread-out model in dimensions $d > 6$, the results of [94] (combined with the above observation from [118] that the triangle is $O(L^{-d})$) give $p_c = |\Omega|^{-1} + O(|\Omega|^{-2})$ as $L \to \infty$. This is improved in [118].[2]

11.3 The Incipient Infinite Cluster

For bond percolation on $\mathbb{Z}^d$, it is believed that there is no percolation at the critical point in all dimensions $d \geq 2$. At present, proofs of this fact are restricted to 2-dimensional and high-dimensional models. The notion of the incipient infinite percolation cluster (IIC) is an attempt to describe the infinite structure that is emerging but not yet materialized at the critical point. Various aspects of the IIC are discussed in [5]. There is currently no existence theory for the IIC that is applicable in general dimensions.

For bond percolation on $\mathbb{Z}^2$, Kesten [142] constructed the IIC as a measure on bond configurations in which the origin is almost surely connected to infinity. He gave two different constructions, both leading to the same measure. One

---

[2] Note that p_c in [118] is $|\Omega|$ times our p_c.

construction is to condition on the event that the origin is connected to infinity, with $p > p_c$, and take the limit $p \to p_c^+$. A second construction is to condition on the event that the origin is connected to the boundary of a box of radius n, with $p = p_c$, and let $n \to \infty$. More recently, Járai [135, 136] has shown that several other definitions of the IIC on $\mathbb{Z}^2$ yield the same measure as Kesten's. These include the inhomogeneous model of [51], and definitions in terms of invasion percolation [49], the largest cluster in a large box [35], and spanning clusters [5]. The incipient infinite cluster is thus a natural and robust object that can be constructed in many different ways, at least in dimension $d = 2$.

In [115], van der Hofstad and Járai used the lace expansion to give two constructions of the IIC above the upper critical dimension. To describe these constructions, we need the following definitions.

Let $\mathcal{F}$ denote the σ-algebra of events. A *cylinder event* is an event that is determined by the occupation status of a finite set of bonds. We denote the algebra of cylinder events by $\mathcal{F}_0$. Then $\mathcal{F}$ is the σ-algebra generated by $\mathcal{F}_0$. For a first definition of the IIC, given $x \in \mathbb{Z}^d$, we begin by defining $\mathbb{P}_x$ by

$$\mathbb{P}_x(E) = \frac{1}{\tau_{p_c}(x)} \mathbb{P}_{p_c}(E \cap \{0 \leftrightarrow x\}) \quad (E \in \mathcal{F}_0). \tag{11.25}$$

We then define $\mathbb{P}_\infty$ by setting

$$\mathbb{P}_\infty(E) = \lim_{x \to \infty} \mathbb{P}_x(E) \quad (E \in \mathcal{F}_0), \tag{11.26}$$

assuming the limit exists. The following theorem shows that, at least in high dimensions, this definition produces a probability measure on $\mathcal{F}$, the IIC measure, under which the origin is almost surely connected to infinity.

Theorem 11.4. *Consider nearest-neighbour or spread-out percolation on $\mathbb{Z}^d$. For the nearest-neighbour model there is a $d_0 \gg 6$, and for the spread-out model there is an $L_0 = L_0(d) \gg 1$, such that the following statements hold for $d \geq d_0$, and for $d > 6$ and $L \geq L_0$, respectively. The limit in (11.26) exists for every cylinder event $E \in \mathcal{F}_0$, independently of the manner in which x goes to infinity. Moreover, $\mathbb{P}_\infty$ extends to a probability measure on the σ-algebra $\mathcal{F}$, and the origin is almost surely connected to infinity under $\mathbb{P}_\infty$.*

The proof of Theorem 11.4 exhibits an explicit cancellation between the numerator and denominator in (11.25). The asymptotic behaviour of the denominator is given by Theorem 11.2 as a multiple of $|x|^{2-d}$. For the numerator, an expansion is developed which is like the expansion for the two-point function, but with notice taken of the fact that the event E occurs. The event E depends on only finitely many bonds, and hence is a local effect in the limit $|x| \to \infty$. Details of the method of proof can be found in [115].

In [115], a second construction of the IIC measure $\mathbb{P}_\infty$ is also given. This construction uses a limit from the *subcritical* side, and proceeds as follows. For $p < p_c$, let

$$\mathbb{Q}_p(E) = \frac{1}{\chi(p)} \sum_{x \in \mathbb{Z}^d} \mathbb{P}_p(E \cap \{0 \leftrightarrow x\}) \quad (E \in \mathcal{F}_0), \tag{11.27}$$

and

$$\mathbb{Q}_{p_c}(E) = \lim_{p \to p_c^-} \mathbb{Q}_p(E) \quad (E \in \mathcal{F}_0), \tag{11.28}$$

assuming the limit exists. It is a theorem of [115] that the limit does exist, and that $\mathbb{Q}_{p_c} = \mathbb{P}_\infty$, under the high-dimension hypotheses of Theorem 11.4.

Once the IIC measure $\mathbb{P}_\infty$ has been constructed, it is natural to ask for its properties, beyond the fact that the origin is almost surely connected to infinity under $\mathbb{P}_\infty$. The following properties are proved in [115] for the spread-out model, and they are expected to be true also for the nearest-neighbour model.

An infinite connected graph is said to have a *single end* if the complement of any finite subgraph contains exactly one infinite connected component. In particular, an infinite connected graph has a single end if any two infinite self-avoiding paths in the set have infinitely many bonds in common. It is proved in [115] that for the spread-out model in dimensions $d > 6$ with L sufficiently large, under $\mathbb{P}_\infty$ any two infinite self-avoiding paths in the cluster of the origin almost surely have infinitely many bonds in common, and hence the cluster has a single end. In addition, the IIC two-point function obeys

$$\mathbb{P}_\infty(0 \leftrightarrow y) \simeq \frac{1}{|y|^{d-4}}, \tag{11.29}$$

whereas

$$\mathbb{P}_\infty(\{0 \leftrightarrow y\} \circ \{y \leftrightarrow \infty\}) \simeq \frac{1}{|y|^{d-2}}. \tag{11.30}$$

Intuitively, these two formulas say that the cluster of the origin is 4-dimensional under $\mathbb{P}_\infty$, whereas the infinite backbone connecting the origin to infinity is 2-dimensional.

It is plausible that the IIC could also be constructed by taking $p = p_c$, conditioning on the event that the origin is in a cluster containing exactly n vertices, and letting $n \to \infty$. This particular construction has not been carried out, but a closely related issue has been studied in [103–105], in high dimensions. In these papers, the cluster of the origin is conditioned to have size n, the lattice spacing is simultaneously rescaled by $n^{-1/4}$, and then the limit $n \to \infty$ is taken. This limit should be regarded as the scaling limit of the IIC. We discuss this issue in more detail in Sect. 16.5, where the scaling limit of the IIC in high dimensions is related to integrated super-Brownian excursion.

Exercise 11.5. The event $\{0 \leftrightarrow y\}$ is not a cylinder event, but it is proved in [115] that (11.26) nevertheless does hold for $E = \{0 \leftrightarrow y\}$. Assuming this, complete the following outline of proof of the upper bound of (11.29).
(a) Apply (9.84) to conclude that

$$\frac{1}{\tau_{p_c}(x)}\mathbb{P}_{p_c}(0 \leftrightarrow y, 0 \leftrightarrow x) \leq \sum_{z\in\mathbb{Z}^d} \tau_{p_c}(z)\tau_{p_c}(y-z)\frac{\tau_{p_c}(x-z)}{\tau_{p_c}(x)}.$$

Using Theorem 11.2, show that the contribution to the above sum due to $|x-z| \leq \frac{1}{2}|x|$ vanishes in the limit $|x| \to \infty$.

(b) Suppose $d > 4$. Prove that if $|f(x)| \leq (|x|+1)^{2-d}$ and $|g(x)| \leq (|x|+1)^{2-d}$ then $|(f*g)(x)| \leq \text{const}\,(|x|+1)^{4-d}$. (See [91, Proposition 1.7] for extensions.)

(c) Analyze the contribution due to $|x-z| > \frac{1}{2}|x|$ by noting that in this case the ratio of two-point functions in the summation is uniformly bounded and converges to 1 as $|x| \to \infty$.

11.4 Percolation on Finite Graphs

It is natural to ask how the percolation phase transition is modified if the infinite graph $\mathbb{Z}^d$ is replaced by a large finite subgraph, such as a d-dimensional box of large radius. Such matters are actually essential to interpret properly the results of computer simulations of percolation, which are necessarily performed on a finite graph. Quite generally, one can inquire about the nature of the percolation phase transition on arbitrary finite graphs. The best understood setting for this question is the complete graph K_V.

11.4.1 The Complete Graph

The complete graph K_V is the graph consisting of V vertices, with an edge joining each of the $\binom{V}{2}$ pairs of vertices. The percolation model is defined as usual: edges are independently occupied with probability p and vacant with probability $1-p$. In the combinatorial literature, percolation on the complete graph is often referred to as random subgraphs of the complete graph, or, more briefly, as the random graph. The substantial theory currently known for the random graph is described in the recent books [26, 134].

For K_V, the susceptibility $\chi(p) = \mathbb{E}_p|C(0)|$ (where 0 denotes an arbitrary vertex) varies continuously between $\chi(0) = 1$ and $\chi(p) = V$. In particular, it is of course never infinite. Similarly, the probability that the origin is an infinite cluster is always zero, and thus the phase transition cannot be characterized by the divergence of χ, or the non-vanishing of θ. Nevertheless, there

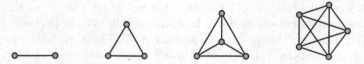

Fig. 11.1. The complete graphs K_2, K_3, K_4, K_5.

$$p = \tfrac{3}{4}\tfrac{1}{\Omega} = .00120 \qquad\qquad p = \tfrac{5}{4}\tfrac{1}{\Omega} = .00200$$

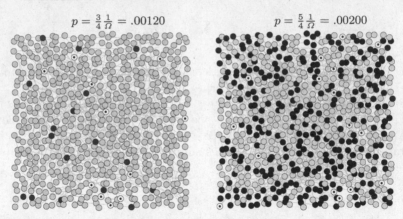

Fig. 11.2. The largest cluster (black) and second largest cluster (dots) in random subgraphs of K_{625}. These clusters have size 17 and 11 on the left, and 284 and 16 on the right. The hundreds of edges in the subgraphs are not clearly shown.

is a phase transition, as suggested by Fig. 11.2. The transition corresponds to an abrupt change in the size $|\mathcal{C}_{\max}|$ of a cluster $\mathcal{C}_{\max}$ of maximal size, as p is varied through the critical value $p_c = \frac{1}{|\Omega|}$, where $|\Omega| = V - 1$ is the degree of the complete graph.

The phase transition on K_V was first studied by Erdős and Rényi [71]. We will say that a sequence of events E_V occurs *with high probability*, denoted w.h.p., if $\mathbb{P}(E_V) \to 1$ as $V \to \infty$. Erdős and Rényi showed that when p is scaled as $(1 + \epsilon)V^{-1}$, there is a phase transition at $\epsilon = 0$ in the sense that w.h.p.

$$|\mathcal{C}_{\max}| \simeq \begin{cases} \log V & \text{if } \epsilon < 0, \\ V^{2/3} & \text{if } \epsilon = 0, \\ V & \text{if } \epsilon > 0. \end{cases} \qquad (11.31)$$

The bounds of (11.31) are valid for fixed ϵ, independent of V.

The results of Erdős and Rényi were substantially extended by Bollobás [25] and Łuczak [156]. In particular, they showed that there is a scaling window of width $V^{-1/3}$ in the sense that if $p = (1 + \Lambda_V V^{-1/3})V^{-1}$, then w.h.p.

$$|\mathcal{C}_{\max}| \begin{cases} \ll V^{2/3} & \text{if } \Lambda_V \to -\infty, \\ \simeq V^{2/3} & \text{if } \Lambda_V \text{ is uniformly bounded in } V, \\ \gg V^{2/3} & \text{if } \Lambda_V \to +\infty. \end{cases} \qquad (11.32)$$

Here, we are using the notation $f(V) \ll g(V)$ to mean that $f(V)/g(V) \to 0$ as $V \to \infty$, while $f(V) \gg g(V)$ means that $f(V)/g(V) \to \infty$ as $V \to \infty$. A great deal more is known, and can be found in [26, 134]. In particular, above the scaling window, taking $\epsilon > 0$ independent of V for simplicity, the second largest cluster has size $\simeq \log V$, much smaller than the largest cluster. This is analogous to the uniqueness of the infinite cluster for supercritical percolation on $\mathbb{Z}^d$ [46].

The complete graph is a particularly simple example of a finite graph. It has a high degree of symmetry, and calculations based on counting arguments can go a long way. What if the complete graph is replaced by a finite graph with more geometrical structure, and less symmetry?

11.4.2 High-Dimensional Transitive Finite Graphs

A detailed study of the phase transition on high-dimensional finite graphs was initiated in [31] and continued in [32, 33, 122, 123]. Let $\mathbb{G} = (\mathbb{V}, \mathbb{B})$ be a graph, where $\mathbb{V}$ is a vertex set of cardinality V, and $\mathbb{B}$ is the edge set. A bijective map $\varphi : \mathbb{V} \to \mathbb{V}$ is called a *graph isomorphism* if $\{\varphi(x), \varphi(y)\} \in \mathbb{B}$ whenever $\{x, y\} \in \mathbb{B}$, and $\mathbb{G}$ is called *vertex transitive* if for every pair of vertices $x, y \in \mathbb{V}$ there is a graph isomorphism φ with $\varphi(x) = y$. In a vertex transitive graph, the graph "looks the same" from each vertex. Transitive graphs are by definition regular, i.e., each vertex has the same degree. We denote the degree by $|\Omega|$. Let $\mathbb{G}$ be any finite connected vertex transitive graph.

In [31], the critical point of $\mathbb{G}$ is defined by analogy with the complete graph, for which it is known that the expected cluster size of any particular vertex, for p inside the scaling window, is of order $V^{1/3}$. Motivated by this fact, the *critical threshold* $p_c = p_c(\mathbb{G}, \lambda)$ of $\mathbb{G}$ was defined to be the unique solution to the equation

$$\chi(p_c) = \lambda V^{1/3}, \tag{11.33}$$

where λ is a positive parameter whose choice is at our disposal. This flexibility in the choice of λ is based on the idea is that as long as λ varies only weakly with V, e.g., if λ is a constant, then the value of $p_c(\mathbb{G}, \lambda)$ should lie in the critical scaling window. In any case, it is necessary to assume that $1 < \lambda V^{1/3} < V$, so that p_c is well defined and $0 < p_c < 1$. To justify the definition, a theorem that a transition really does take place at p_c is required.

The definition (11.33) is appropriate for graphs that obey mean-field behavior, which we expect only for graphs that are in some sense "high-dimensional." In particular, it is argued in [31] that (11.33) should be replaced by $\chi(p_c) = \lambda V^{(\delta-1)/(\delta+1)}$, where δ is the critical exponent for the magnetization (9.18), when $\mathbb{G}$ is a finite periodic approximation to $\mathbb{Z}^d$ with $d < 6$. Since the mean-field value of δ is 2, this is consistent with the definition (11.33).

In [31], the triangle condition is modified to apply to finite graphs, as follows. Let

$$\nabla_p(x, y) = \sum_{u, v \in V} \tau_p(x, u) \tau_p(u, v) \tau_p(v, y) \tag{11.34}$$

denote the (open) triangle diagram. Then the triangle condition is the statement that

$$\max_{x, y \in \mathbb{V}} \left[\nabla_{p_c(\mathbb{G}, \lambda)}(x, y) - \delta_{x,y} \right] \leq a_0, \tag{11.35}$$

where a_0 is a sufficiently small number. This is a natural analogue of the condition that $\bar{\mathsf{T}}(p_c)$ of (9.54) be small for $\mathbb{Z}^d$. Since $\sum_y \nabla_p(x, y) = \chi(p)^3$, the triangle condition implies that $\lambda^3 \leq V^{-1} + a_0$, which requires small λ.

It is proved in [31] that if the triangle condition holds, then many aspects of the phase transition for $\mathbb{G}$ resemble their counterparts for the complete graph. In particular, $|\mathcal{C}_{\max}| \simeq V^{2/3}$ w.h.p. inside a scaling window of size $V^{-1/3}$ around $p_c(\mathbb{G}, \lambda)$. Further consequences of the triangle condition are discussed below. The method of proof is based in part on an adaptation of the differential inequality methods of Chap. 9, and in part on ideas from [35].

To apply the results of [31], it is necessary to establish the triangle condition. This is the subject of [32], where the expansion of Chap. 10 is used to prove the triangle condition for various high-dimensional graphs. These graphs, denoted $\mathbb{T}_{r,d}$, all have vertex set of the form $\mathbb{V} = \{0, 1, \ldots, r-1\}^d$, and edge set such that $\{0, x\}$ is an edge if and only if $\{y, y \pm x\}$ is an edge for all vertices y (translation and reflection symmetry). The addition on $\mathbb{V}$ is componentwise addition modulo r, corresponding to periodic boundary conditions. It is shown that for such graphs a random walk version of the triangle condition, closely related to (10.59), implies the percolation triangle condition. For many graphs, the random walk triangle condition is relatively easy to establish. Full details can be found in [32]. Among the examples for which the triangle condition is proven in [32] are:

1. the spread-out torus, with $\mathbb{B} = \{\{x, y\} : 0 < \max_{i=1,\ldots,d} |x_i - y_i| \leq L\}$, with $d \geq 7$ fixed, L large and fixed, in the limit $r \to \infty$. Here $V = r^d$ and $|\Omega| = (2L+1)^d - 1$.
2. the d-cube $\mathbb{Q}_d$, with $\mathbb{V} = \{0, 1\}^d$, and with an edge joining distinct vertices that agree in all but one coordinate. Here $V = 2^d$ and $|\Omega| = d$.

The following is a partial list of results from [32] for a finite connected vertex transitive graph that is assumed to obey the triangle condition. The results include upper and lower bounds below and within a scaling window of width $V^{-1/3}$, and upper bounds (but not lower bounds) above the scaling window. For the particular case $\mathbb{G} = \mathbb{Q}_d$, lower bounds above the scaling window will be discussed later.

Asymptotic behaviour of the critical value p_c. This is given by

$$p_c(\mathbb{G}; \lambda) = \frac{1}{|\Omega|}\left[1 + O(|\Omega|^{-1}) + O(\lambda^{-1}V^{-1/3})\right]. \qquad (11.36)$$

Asymptotics below the window. Let $p = p_c - |\Omega|^{-1}\epsilon$ with $\epsilon \geq 0$. If $\epsilon\lambda V^{1/3} \to \infty$ as $V \to \infty$, then as $V \to \infty$,

$$\chi(p) = \frac{1}{\epsilon}\left[1 + O(|\Omega|^{-1}) + O((\epsilon\lambda V^{1/3})^{-1})\right]. \qquad (11.37)$$

Comparing with (9.12), this is a statement that $\gamma = 1$. In addition, for all $\epsilon \geq 0$,

$$\mathbb{P}_p\left(|\mathcal{C}_{\max}| \leq 2\chi(p)^2 \log(V/\chi(p)^3)\right) \geq 1 - \frac{\sqrt{e}}{[2\log(V/\chi(p)^3)]^{3/2}}. \qquad (11.38)$$

In particular, if $\epsilon \lambda V^{1/3} \to \infty$, then $|\mathcal{C}_{max}| \leq O(\epsilon^{-2} \log V)$ w.h.p., and it is also shown in [31] that in this case $|\mathcal{C}_{max}| \geq \mathrm{const}\, \epsilon^{-2}$ w.h.p.

Asymptotics inside the window. Fix $\Lambda < \infty$. There exist constants b_i such that the following hold for all $p = p_c + |\Omega|^{-1}\epsilon$ with $|\epsilon| \leq \Lambda V^{-1/3}$, i.e., inside a scaling window of width proportional to $V^{-1/3}$. If $k \leq b_1 V^{2/3}$, then

$$\frac{b_2}{\sqrt{k}} \leq \mathbb{P}_p(|C(0)| \geq k) \leq \frac{b_3}{\sqrt{k}}. \tag{11.39}$$

The bounds (11.39) are a statement that $\delta = 2$ (cf. (9.19)). If $\omega \geq 1$, then

$$\mathbb{P}_p\left(\omega^{-1}V^{2/3} \leq |\mathcal{C}_{max}| \leq \omega V^{2/3}\right) \geq 1 - \frac{b_6}{\omega}. \tag{11.40}$$

Finally,

$$b_7 V^{1/3} \leq \chi(p) \leq b_8 V^{1/3}. \tag{11.41}$$

In the above statements, the constants b_2 and b_3 can be chosen independent of λ and Λ, the constant b_8 depends on Λ and not on λ, and the constants b_1, b_6 and b_7 depend on both λ and Λ.

Upper bounds above the window. For the supercritical phase, let $p = p_c + \epsilon|\Omega|^{-1}$ with $\epsilon \geq 0$. Then for all $\omega > 0$,

$$\mathbb{P}_p\left(|\mathcal{C}_{max}| \leq \omega(V^{2/3} + \epsilon V)\right) \geq 1 - \frac{21}{\omega}, \tag{11.42}$$

and

$$\chi(p) \leq 81(V^{1/3} + \epsilon^2 V). \tag{11.43}$$

Note that ϵV dominates $V^{2/3}$ and $\epsilon^2 V$ dominates $V^{1/3}$ when $\epsilon \geq \mathrm{const}\, V^{-1/3}$. The bounds (11.42)–(11.43) provide upper bounds on the size of clusters in the supercritical phase. To see that a phase transition occurs at p_c, one wants a *lower* bound. The results of [33] provide such a lower bound for the d-cube.

11.4.3 The d-Cube

For the d-cube, it is proved in [33] that there are strictly positive constants c_0, c_1, c_2 such that the following holds as $d \to \infty$ for all d-independent λ with $0 < \lambda \leq c_0$ and all $p = p_c + \epsilon d^{-1}$ with $e^{-c_1 d^{1/3}} \leq \epsilon \leq 1$:

$$|\mathcal{C}_{max}| \geq c_2 \epsilon 2^d \quad \text{w.h.p.}, \tag{11.44}$$

$$\chi(p) \geq (c_2\epsilon)^2 2^d. \tag{11.45}$$

These bounds are complementary to (11.42)–(11.43). This leaves only the tantalizingly small interval $2^{-d/3} \ll \epsilon \leq e^{-c_1 d^{1/3}}$ where lower bounds are lacking.

For the d-cube, (11.36) states that if λ is chosen such that $\lambda^{-1} 2^{-d/3} = O(d^{-1})$, then $p_c(\mathbb{Q}_d, \lambda) = \frac{1}{d} + O(\frac{1}{d^2})$. This result has been extended in [122] to show that there are rational numbers b_i $(i \geq 1)$ such that for all positive integers M, all $c, c' > 0$, and all p for which $\chi(p) \in [cd^M, c'd^{-2M}2^n]$,

$$p = \sum_{i=1}^{M} b_i d^{-i} + O(d^{-M-1}), \qquad (11.46)$$

where the constant in the error term depends only on c, c', M. In particular, the above holds for p such that $\chi(p) = \lambda 2^{d/3}$ for *any* fixed $\lambda > 0$. Thus

$$p_c(\mathbb{Q}_d, \lambda) \sim \sum_{i=1}^{\infty} b_i d^{-i} \qquad (11.47)$$

is an asymptotic expansion for any fixed λ (in fact, the asymptotic expansion is shown in [122] to hold even with substantial d-dependence of λ). The asymptotic expansion is conjectured to be divergent. It follows from (11.36) that $b_1 = 1$, and it is shown in [123] that $b_2 = 1$ and $b_3 = \frac{7}{2}$. Thus,

$$p_c(\mathbb{Q}_d, \lambda) = \frac{1}{d} + \frac{1}{d^2} + \frac{7}{2}\frac{1}{d^3} + O\left(\frac{1}{d^4}\right), \qquad (11.48)$$

and the first three coefficients in the expansion agree with the first three coefficients in the expansion (11.19) for the critical value for $\mathbb{Z}^d$. It is conjectured that the expansions differ in the fourth term, i.e., that $a_4 \neq b_4$.

These results for the d-cube extend results of [13], who identified $\frac{1}{d}$ as a threshold by studying the largest cluster for $p = \frac{1}{d}(1 + \epsilon)$ with ϵ independent of d, and [28], who obtained results for ϵ depending on d but not as small as $\frac{1}{d}$.

11.4.4 Large Finite Boxes in $\mathbb{Z}^d$

For percolation in a box of radius r in $\mathbb{Z}^d$, a study of the phase transition was undertaken in [35]. The analysis of [35] is based on certain scaling and hyperscaling hypotheses which are conjectured to be valid in dimensions $d = 3, 4, 5, 6$ and are known to hold in dimension 2. It provides a partial description of finite-size scaling for percolation below the upper critical dimension. In dimensions $d > 6$, hyperscaling does not hold, and a different theory is needed.

In [5], Aizenman considered the size of the largest cluster inside a box of radius r in $\mathbb{Z}^d$ for $d > 6$, with the *bulk* boundary condition. Under the bulk boundary condition, vertices inside the box are considered to be connected if there is an occupied path that joins them in the infinite lattice $\mathbb{Z}^d$. In particular, connections are permitted to exit the box. It was shown in [5] that for $d > 6$, roughly speaking, at p_c the largest cluster in the box has size r^4 and there are of the order of r^{d-6} clusters of this size, under the assumption that $\tau_{p_c}(x) \simeq |x|^{2-d}$. This last hypothesis is supplied by Theorem 11.2, under the usual high-d assumptions. The fact that the largest cluster has size r^4 is another indication of the 4-dimensional character of the IIC, when $d > 6$.

In terms of the volume $V = r^d$ of the box, the largest cluster at the critical value $p_c(d)$ for $\mathbb{Z}^d$, under the bulk boundary condition, thus has size of order $r^4 = V^{4/d}$, for $d > 6$. The question was raised in [5] whether this would change to $r^{2d/3} = V^{2/3}$ if the periodic boundary condition is used instead of the bulk boundary condition. According to (11.40), for the spread-out model in dimensions $d > 6$ with periodic boundary conditions, if p is within a scaling window centred at $p_c(\mathbb{T}_{r,d}; \lambda)$ (with λ small and constant) and of width proportional to $V^{-1/3}$, then the largest cluster is of size $V^{2/3}$. As was pointed out in [31], an affirmative answer to the question would follow if it could be proved that the critical value $p_c(d)$ of the infinite lattice $\mathbb{Z}^d$ is within this scaling window. It is an open problem to do so.

Oriented Percolation

Oriented percolation is a percolation model in which bonds and connections are directed. The model enjoys a Markov property not present in ordinary percolation, and this simplifies some aspects of the problem. In this chapter, we survey some of the results that have been obtained for the critical behaviour of oriented percolation, above its upper critical dimension $d = 4$. The expansions used to obtain these results will be discussed in Chap. 13.

12.1 The Phase Transition

Let Ω denote either the nearest-neighbour or spread-out sets defined in (1.1)–(1.2). Let $\mathbb{Z}^d \times \mathbb{Z}_+$ denote the graph with vertex set consisting of pairs (x, n) with $x \in \mathbb{Z}^d$ and n a non-negative integer, and with directed edges given by ordered pairs $((x, n), (y, n + 1))$ of vertices whose time variables differ by 1 and whose space variables obey $y - x \in \Omega$. As usual, we refer to edges as bonds. Bonds are independently occupied with probability p and vacant with probability $1 - p$. We write $\{(x, m) \to (y, n)\}$ to denote the event that there is a directed path from (x, m) to (y, n) consisting of occupied bonds, i.e., there is a sequence of occupied bonds $((u_{i-1}, i - 1), (u_i, i))$, for $i = m + 1, \ldots, n$, such that $(u_m, m) = (x, m)$ and $(u_n, n) = (y, n)$.

Let

$$C(0,0) = \{(x, n) : (0, 0) \to (x, n)\} \tag{12.1}$$

(see Fig. 12.1). The susceptibility is defined by

$$\chi(p) = \mathbb{E}_p |C(0, 0)| \tag{12.2}$$

and the percolation probability is defined by

$$\theta(p) = \mathbb{P}_p(|C(0, 0)| = \infty). \tag{12.3}$$

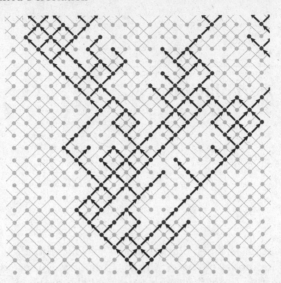

Fig. 12.1. The connected cluster $C(0,0)$ of the origin for oriented percolation on $\mathbb{Z} \times \mathbb{Z}_+$. Here $p = 0.7$ is a little above $p_c \approx 0.645$. For the nearest-neighbour model depicted, the lattice decomposes into two noncommunicating lattices. This will not be the case for the spread-out model when $L \geq 2$.

The phase transition for oriented percolation occurs at a point that is characterized exactly as it is for ordinary percolation, namely [7, 161, 162]

$$p_c = \sup\{p : \theta(p) = 0\} = \sup\{p : \chi(p) < \infty\}. \tag{12.4}$$

For oriented percolation it has been proved that $\theta(p_c) = 0$ for *all* dimensions $d \geq 1$ [22].

The magnetization is defined, for $\gamma \in [0, 1]$, by[1]

$$M(p, \gamma) = 1 - \sum_{n=1}^{\infty}(1 - \gamma)^n \mathbb{P}_p(|C(0,0)| = n). \tag{12.5}$$

Critical exponents analogous to those in Sect. 9.1 can also be defined for oriented percolation. In particular, writing c to denote a positive constant whose value is unimportant and may change from line to line, the following power laws are believed to hold:

---

[1] This γ should not be confused with the critical exponent in (12.6) which is denoted by the same Greek letter.

$$\chi(p) \sim c(p_c - p)^{-\gamma} \quad \text{as } p \to p_c^-, \tag{12.6}$$

$$M(p_c, \gamma) \sim c\gamma^{1/\delta} \quad \text{as } \gamma \to 0, \tag{12.7}$$

$$\theta(p) \sim c(p - p_c)^\beta \quad \text{as } p \to p_c^+, \tag{12.8}$$

$$\frac{\mathbb{E}_p(|C(0,0)|^{m+1})}{\mathbb{E}_p(|C(0,0)|^m)} \sim c(p_c - p)^{-\Delta} \quad \text{as } p \to p_c^-, \text{ for } m = 1, 2, \ldots, \tag{12.9}$$

for some amplitudes c and universal critical exponents γ, δ, β, Δ.

It was first predicted in [172] that the upper critical dimension for percolation is $d = 4$, or, to emphasize the additional time dimension, $d + 1 = 4 + 1$. This is now a theorem, as the hyperscaling inequalities of [180] show that the upper critical dimension is not less than $4 + 1$, and the results below show it is not greater than $4 + 1$.

12.2 The Infrared Bound and the Triangle Condition

The infrared bound and the triangle condition play important roles for the critical behaviour of oriented percolation, just as they do for ordinary percolation. In this section, we discuss this.

The *two-point function* $\tau_{p,n}(x)$ is defined by

$$\tau_{p,n}(x) = \mathbb{P}_p((0,0) \to (x,n)). \tag{12.10}$$

When p is understood from the context, we often drop the subscript p, writing simply $\tau_n(x)$.

By definition,

$$\hat{\tau}_{p,n}(0) = \sum_x \tau_{p,n}(x) = \mathbb{E}_p|C(0,0) \cap (\mathbb{Z}^d \times \{n\})|. \tag{12.11}$$

Thus $\hat{\tau}_{p,n}(0)$ is equal to the expected number of vertices that the origin is connected to at time n. This quantity is submuliplicative, in the sense that

$$\hat{\tau}_{p,m+n}(0) \le \hat{\tau}_{p,m}(0)\hat{\tau}_{p,n}(0). \tag{12.12}$$

To see this, note that a connection from the origin to $(x, m + n)$ implies the existence of a vertex (y, m) such that $(0,0) \to (y,m)$ and $(y,m) \to (x, m+n)$. These two events are independent—this is the Markov property of oriented percolation—so

$$\tau_{p,m+n}(x) \le \sum_y \tau_{p,m}(y)\tau_{p,n}(x-y), \tag{12.13}$$

and the desired inequality follows by summation over x. It then follows from (12.12) that the limit

$$\log r(p) = - \lim_{n \to \infty} \frac{1}{n} \log \hat{\tau}_{p,n}(0) \tag{12.14}$$

exists. By the monotonicity of $\hat{\tau}_{p,n}(0)$ in p, $r(p)$ is non-increasing in p. Various arguments lead to the conclusion that $r(p)$ is finite and positive for $p \leq p_c$, with $r(p_c) = 1$ (see [12,167]).

We define a generating function

$$\hat{t}_p(k,z) = \sum_{n=0}^{\infty} \hat{\tau}_{p,n}(k) z^n, \qquad (12.15)$$

which has radius of convergence $r(p)$ when $k = 0$, by definition of $r(p)$. This is a Fourier–Laplace transform of the two-point function. The infrared bound is an upper bound on $|\hat{t}_p(k,z)|$, given in the following theorem of Nguyen and Yang [167]. The form of the infrared bound can be predicted by comparison with the random walk analogue of (12.15), which is

$$\sum_{n=0}^{\infty} \hat{D}(k)^n z^n = \frac{1}{1 - z\hat{D}(k)} = \frac{1}{[1 - \hat{D}(k)] + (1 - z)\hat{D}(k)}, \qquad (12.16)$$

with D given by (1.10). The first term in the denominator is quadratic in k, for small k, while the second is linear in $1 - z$. If we take $z = e^{i\theta}$ with $\theta \in [-\pi, \pi]$, an upper bound on the absolute value of (12.16) is const $[|k|^2 + |\theta|]^{-1}$.

Theorem 12.1. *There exist d_0 and L_0 such that the following statement holds for the nearest-neighbour model in dimensions $d \geq d_0$ and for the spread-out model in dimensions $d > 4$ for $L \geq L_0$. Let $p \in (0, p_c]$, $k \in [-\pi, \pi]^d$, $z \in \mathbb{C}$ with $|z| < r(p)$. Then*

$$|\hat{t}_p(k,z)| \leq \frac{C_1}{|k|^2 + |r(p) - z|}, \qquad (12.17)$$

where C_1 is a constant that may depend on d and L.

In particular, for $p < p_c$ and $z = e^{i\theta}$, the denominator of the infrared bound can be replaced by $|k|^2 + |\theta|$ in an upper bound. The linear term $|\theta|$, which is a symptom of the orientation in the direction of increasing time, will have an impact on the upper critical dimension.

The proof of Theorem 12.1 uses the lace expansion discussed below in Sect. 13.2, with a convergence proof based on the bootstrap Lemma 5.9. An important consequence is an oriented percolation version of the triangle condition, which we discuss next.

The derivation of differential inequalities involving the triangle condition was given in [20] in a very general setting that allows for oriented bonds, and the results of Chap. 9 have counterparts also for oriented percolation. In brief, an appropriate statement of the triangle condition implies existence of the critical exponents γ, δ, β, Δ, in the sense of $\simeq$, with the mean-field values $\gamma = \beta = 1$ and $\delta = \Delta = 2$. The form of the triangle condition is modified by the orientation, as follows.

Let

$$\nabla_p(x,n) = \sum_{(y,l)} \sum_{(w,m)} \tau_{p,l}(y)\tau_{p,m-l}(w-y)\tau_{p,m-n}(w-x). \qquad (12.18)$$

Using $|(x,n)|$ to denote the Euclidean norm of (x,n) in $\mathbb{Z}^{d+1}$, the triangle condition is the statement that

$$\lim_{R\to\infty} \sup\{\nabla_{p_c}(x,n) : |(x,n)| \geq R\} = 0. \qquad (12.19)$$

Exercise 12.2. Complete the following outline to prove that the triangle condition follows from (12.17), when $d > 4$. Consider $p < p_c$ and take $z = e^{i\theta}$. Show that (12.18) can be rewritten as

$$\nabla_p(x,n) = \int_{-\pi}^{\pi} \frac{d\theta}{2\pi} \int_{[-\pi,\pi]^d} \frac{d^dk}{(2\pi)^d} \hat{t}_p(k,e^{i\theta})^2 \hat{t}_p(k,e^{-i\theta}) e^{-ik\cdot x} e^{-i\theta n}. \qquad (12.20)$$

The infrared bound implies that

$$|\hat{t}_p(k,e^{\pm i\theta})| \leq \frac{c}{|k|^2 + |\theta|}. \qquad (12.21)$$

Conclude from the Riemann–Lebesgue lemma that the triangle condition holds if $d > 4$.

In [167], Nguyen and Yang used the generating function methods of [97] to extend their results to obtain certain statements in terms of asymptotic formulas, instead of upper and lower bounds with different constants. For the important particular case $p = p_c$, and for the spread-out model, the following theorem from [121] extends the results of [167] (for results when $p < p_c$, see [167]). The theorem is proved using the general inductive approach to the lace expansion [120]. Versions of parts (a) and (b) of the theorem are proved in [167], with weaker error estimates, and also for the nearest-neighbour model for $d \geq d_0$.

Theorem 12.3. *Consider the spread-out model of oriented percolation. Let $d > 4$, $p = p_c$, and $\delta \in (0, 1 \wedge \frac{d-4}{2})$. There is an $L_0 = L_0(d)$ such that for $L \geq L_0$ there exist positive constants v and A (depending on d and L), and C_1, C_2 (depending only on d), such that the following statements hold as $n \to \infty$:*
(a)

$$\hat{\tau}_{p_c,n}(k/\sqrt{vn}) = Ae^{-|k|^2/2d}[1 + O(|k|^2n^{-\delta}) + O(n^{-(d-4)/2})], \qquad (12.22)$$

(b)

$$\frac{1}{\hat{\tau}_{p_c,n}(0)} \sum_x |x|^2 \tau_{p_c,n}(x) = vn[1 + O(n^{-\delta})], \qquad (12.23)$$

(c)

$$C_1 L^{-d} n^{-d/2} \leq \sup_{x\in\mathbb{Z}^d} \tau_{p_c,n}(x) \leq C_2 L^{-d} n^{-d/2}, \qquad (12.24)$$

with the error estimate in (a) uniform in $k \in \mathbb{R}^d$ with $|k|^2(\log n)^{-1}$ sufficiently small.

Theorem 12.3 is the $r = 2$ case of a family of results which give the asymptotics of the critical oriented percolation r-point functions for all $r \geq 2$. The r-point functions for $r \geq 3$ will be discussed in Sect. 17.3.

Theorem 12.3(a) shows that the expected number of particles $\hat{\tau}_n(0)$ to which the origin is connected at time n converges to a nonzero finite constant A as $n \to \infty$, when $p = p_c$. In contrast, for $p < p_c$ in general dimensions, $\hat{\tau}_n(0)$ decays exponentially to zero, while for $p > p_c$ the limit is infinite. (See [22, 86] for the relevant shape theorem when $p > p_c$.)

Theorem 12.3(b) shows that at p_c the length scale at time n is of order $n^{1/2}$.

The bounds of Theorem 12.3(c) are consistent with a local central limit theorem. A version of the local central limit theorem similar to Theorem 6.5 applies also to oriented percolation [120, 121].

An extension of Theorem 12.3 to dimensions $d \leq 4$ is discussed below Theorem 14.3, for the case of an oriented percolation model whose range L is unbounded.

Exercise 12.4. Complete the following outline to derive the triangle condition from Theorem 12.3(a, c).
(a) Argue that if $|(x, n)| \geq R$ in (12.18), then the summation index m can be taken to be at least a multiple of R in the sum in (12.18). Thus it suffices to prove that

$$\lim_{R \to \infty} \sum_{(y,l)} \sum_{(w,m):m \geq cR} \tau_{p_c,l}(y)\tau_{p_c,m-l}(w - y)\tau_{p_c,m-n}(w - x) = 0. \qquad (12.25)$$

(b) Obtain this sufficient condition by considering separately the contributions to the sum due to $l \leq m/2$ and $l > m/2$. Use $\|\tau_{p_c,n}\|_1 \leq O(1)$ and $\|\tau_{p_c,n}\|_\infty \leq O(n^{-d/2})$, noting that the latter is most effectively applied when n is large.

12.3 The Critical Value

It was observed in [167] that the estimates of [166] imply that the critical value for nearest-neighbour oriented percolation is given asymptotically as

$$p_c = \frac{1}{2d} + O\left(\frac{1}{(2d)^2}\right). \qquad (12.26)$$

Other methods [57] have produced the stronger result that

$$d^{-1} + \frac{1}{2}d^{-3} + o(d^{-3}) \leq p_c \leq d^{-1} + d^{-3} + O(d^{-4}), \qquad (12.27)$$

for a version of oriented percolation in which the forward degree of a vertex is d rather than $2d$. For the spread-out model, it is shown in [121] that $p_c = |\Omega|^{-1} + O(|\Omega|^{-2})$ when $d > 4$, and this is improved in [118].[2]

---

[2] Note that p_c in [118] is $|\Omega|$ times our p_c.

12.4 The Incipient Infinite Cluster

The incipient infinite cluster for ordinary percolation was discussed in Sect. 11.3. In this section, we state existence results for the IIC for spread-out oriented percolation in dimensions $d > 4$, and indicate some properties of the IIC.

We write $\mathcal{F}$ for the σ-algebra of events, and denote the algebra of cylinder events (those determined by the occupation status of a finite set of bonds) by $\mathcal{F}_0$. Then $\mathcal{F}$ is the σ-algebra generated by $\mathcal{F}_0$. The following are four possible definitions of the IIC. In each definition, we assume initially that $E \in \mathcal{F}_0$.

1. Define $\mathbb{P}_n$ by

$$\mathbb{P}_n(E) = \frac{1}{\hat{\tau}_n(0)} \sum_{x \in \mathbb{Z}^d} \mathbb{P}(E \cap \{(0,0) \to (x,n)\}), \qquad (12.28)$$

and, assuming the limit exists, define $\mathbb{P}_\infty$ by setting

$$\mathbb{P}_\infty(E) = \lim_{n \to \infty} \mathbb{P}_n(E). \qquad (12.29)$$

2. Let $S_n = \{\exists x \in \mathbb{Z}^d \text{ such that } (0,0) \to (x,n)\}$ denote the event that the cluster of the origin survives to time n. Define $\mathbb{Q}_n$ by

$$\mathbb{Q}_n(E) = \mathbb{P}(E|S_n), \qquad (12.30)$$

and, assuming the limit exists, define $\mathbb{Q}_\infty$ by setting

$$\mathbb{Q}_\infty(E) = \lim_{n \to \infty} \mathbb{Q}_n(E). \qquad (12.31)$$

3. Fix any $x \in \mathbb{Z}^d$, define $\mathbb{P}_n^{(x)}$ by

$$\mathbb{P}_n^{(x)}(E) = \mathbb{P}(E|\{(0,0) \to (x,n)\}), \qquad (12.32)$$

and, assuming the limit exists, define $\mathbb{P}_\infty^{(x)}$ by

$$\mathbb{P}_\infty^{(x)}(E) = \lim_{n \to \infty} \mathbb{P}_n^{(x)}(E). \qquad (12.33)$$

4. For $p < p_c$, let

$$\mathbb{K}_p(E) = \frac{1}{\chi(p)} \sum_{x \in \mathbb{Z}^d} \sum_{n=0}^{\infty} \mathbb{P}_p(E \cap \{(0,0) \to (x,n)\}), \qquad (12.34)$$

and, assuming the limit exists, define $\mathbb{K}_\infty$ by

$$\mathbb{K}_\infty(E) = \lim_{p \to p_c^-} \mathbb{K}_p(E). \qquad (12.35)$$

It is natural to conjecture that the above limits exist in all dimensions, and that, moreover, for all x,

$$\mathbb{P}_\infty = \mathbb{Q}_\infty = \mathbb{P}_\infty^{(x)} = \mathbb{K}_\infty. \tag{12.36}$$

The following theorem shows that for sufficiently spread-out oriented percolation in dimensions $d > 4$, the limits $\mathbb{P}_\infty$ and $\mathbb{K}_\infty$ exist and are equal. In addition, $\mathbb{Q}_\infty$ exists and equals $\mathbb{P}_\infty = \mathbb{K}_\infty$, using the fact that the critical survival probability

$$\theta_n = \mathbb{P}(S_n) \tag{12.37}$$

obeys

$$\theta_n \sim \frac{1}{Bn} \tag{12.38}$$

for some positive constant B. It is a classical fact that the probability that a critical branching process survives for at least n generations is asymptotically a constant multiple of n^{-1}. The asymptotic formula (12.38), which is proved in [111,112], shows that this mean-field behaviour applies also to sufficiently spread-out oriented percolation when $d > 4$. The proof of (12.38) uses a *point-to-plane* expansion rather than the usual point-to-point expansion. The theorem does not make any statement about the limit $\mathbb{P}_\infty^{(x)}$.

Theorem 12.5. *Let $d + 1 > 4 + 1$ and $p = p_c$. There is an $L_0 = L_0(d)$ such that for $L \geq L_0$, the limits $\mathbb{P}_\infty$ and $\mathbb{K}_\infty$ exist and are equal for every cylinder event $E \in \mathcal{F}_0$. When we also apply (12.38), the limit $\mathbb{Q}_\infty$ exists and equals $\mathbb{P}_\infty = \mathbb{K}_\infty$. Moreover, the limit $\mathbb{P}_\infty$ (and hence also $\mathbb{Q}_\infty$ and $\mathbb{K}_\infty$) extends to a probability measure on the σ-algebra $\mathcal{F}$, and the origin is almost surely connected to infinity under $\mathbb{P}_\infty$.*

The proof of Theorem 12.5 can be found in [114], except for the statement concerning $\mathbb{K}_\infty$, which is proved in [115]. The proof is based on an extension of the expansion methods discussed in Chap. 13.

Next, we discuss some properties of the IIC measure $\mathbb{P}_\infty$, which are proved in [114]. The Hausdorff dimension of the connected cluster of the origin under the IIC is predicted to equal 4 almost surely, for $d + 1 > 4 + 1$. The following theorem provides a weaker statement indicating a 4-dimensional aspect to the IIC.

Theorem 12.6. *Let $d + 1 > 4 + 1$ and $p = p_c$. There exists $L_0 = L_0(d)$ such that for $L \geq L_0$,*

$$\mathbb{E}_\infty \left[|\{(y, m) \in C(0, 0) : |y| \leq R\}| \right] \simeq R^4, \tag{12.39}$$

where the expectation is with respect to the IIC measure $\mathbb{P}_\infty$, and the absolute value signs represent cardinality of the set.

Let

$$N_m = |\{y \in \mathbb{Z}^d : (0,0) \leftrightarrow (y,m)\}| \tag{12.40}$$

denote the number of vertices at time m to which the origin is connected. The following theorems give the limiting distribution of N_m under $\mathbb{P}_\infty$ and $\mathbb{Q}_m$. The constants A and B in the theorems are the constants of Theorem 12.3(a) and (12.38), while the constant V is the vertex factor which appears in the scaling limit of the 3-point function (see Theorem 17.5 below). Also, we recall that a *size-biased* exponential random variable with parameter λ has density

$$f(x) = \lambda^2 x e^{-\lambda x} \qquad (x \geq 0). \tag{12.41}$$

Theorem 12.7. *(a) Let $d+1 > 4+1$ and $p = p_c$. There is an $L_0 = L_0(d)$ such that for $L \geq L_0$ and for $l = 1, 2, \ldots$,*

$$\lim_{m\to\infty} \mathbb{E}_\infty\left[\left(\frac{N_m}{m}\right)^l\right] = \left(\frac{A^2 V}{2}\right)^l (l+1)!. \tag{12.42}$$

Consequently, under $\mathbb{P}_\infty$, $m^{-1}N_m$ converges weakly to a size-biased exponential random variable with parameter $\lambda = 2/(A^2 V)$.
(b) Assume in addition that (12.38) holds. Then $B = AV/2$, and for $l = 1, 2, \ldots$,

$$\lim_{m\to\infty} \mathbb{E}_{\mathbb{Q}_m}\left[\left(\frac{N_m}{m}\right)^l\right] = \left(\frac{A^2 V}{2}\right)^l l!. \tag{12.43}$$

Consequently, under $\mathbb{Q}_m$, $m^{-1}N_m$ converges weakly to an exponential random variable with parameter $\lambda = 2/(A^2 V)$.

Exercise 12.8. Show that (12.42) and (12.43) imply the weak convergence statements made in Theorem 12.7.

Under the assumption that (12.38) holds, it follows from Theorems 12.7 that $m^{-1}N_m$ converges to a size-biased exponential random variable under $\mathbb{Q}_\infty = \mathbb{P}_\infty$, and to an exponential random variable under $\mathbb{Q}_m$. This is similar to the situation for critical branching processes, for which the size-biased exponential distribution occurs when the branching random walk is conditioned to survive to infinite time, and the exponential distribution occurs when the branching random walk is conditioned to survive until time m.

The IIC r-point functions and the scaling limit of the IIC will be discussed in Sect. 17.3. Exercise 17.6 concerns the proof of Theorem 12.7(a).

13

Expansions for Oriented Percolation

The expansion of Sect. 10.1 applies directly to oriented percolation, and is sometimes the expansion of choice for oriented percolation [121]. The translation of the expansion of Sect. 10.1 to the oriented setting is described in Sect. 13.3.

However, it is also possible to exploit the Markov property of oriented percolation to give alternate, simpler versions of the expansion. A derivation using laces and resummation, first carried out by Nguyen and Yang [166, 208], is described in Sect. 13.2. Later, a particularly simple derivation of the expansion, based on inclusion-exclusion, was given by Sakai [179]. Sakai's method is described in Sect. 13.1.

Each expansion gives rise to a recursion equation

$$\tau_n(x) = (p|\Omega|D * \tau_{n-1})(x) + \sum_{m=2}^{n-1} (\pi_m * p|\Omega|D * \tau_{n-1-m})(x) + \pi_n(x), \quad (13.1)$$

valid for $n \geq 1$. The empty sum is zero for $n \leq 2$, and substituting $n = 1$ shows that $\pi_1(x) = 0$. The function D is given by (1.10). The identity (13.1) can be used as a recursive definition of $\pi_m(x)$, by isolating the term $\pi_n(x)$ on the right hand side and noting that this expresses it in terms of the two-point function and π_m with $m < n$. Therefore, $\pi_n(x)$ is uniquely determined by (13.1). It follows that all methods for deriving (13.1) must yield exactly the same quantities $\pi_m(x)$.

On the other hand, there are differences between the representations for $\pi_m(x)$ that are produced by the different expansion methods. Each method gives a representation of π as an alternating sum of terms $\pi^{(N)}$. For the percolation expansion of Sect. 10.1 applied to oriented percolation, the individual terms $\pi^{(N)}$ in the sum are again given by nested expansions as in (10.28). On the other hand, the other two expansions each produce identical representations for $\pi^{(N)}$ by a *single* expectation, which is conceptually simpler. In all cases, the diagrammatic estimates for $\pi^{(N)}$ are ultimately very similar, and these are briefly discussed in Sect. 13.5.

13.1 Inclusion-Exclusion

In this section, we derive the expansion (13.1) using the method of Sakai [179].

We write $\{(x, n) \Rightarrow (y, m)\}$ to denote the event that there are at least two bond-disjoint connections from (x, n) to (y, m), or that $(x, n) = (y, m)$. As in Definition 9.13(c), given a bond b we define $\tilde{C}^b(x, n)$ to be the random set of vertices to which (x, n) remains connected after the bond b is made vacant.

As in (10.8), the first step of the expansion is to partition the event $\{(0, 0) \rightarrow (x, n)\}$ according to whether or not there is a double connection. This gives

$$\tau_n(x) = \mathbb{P}((0, 0) \Rightarrow (x, n)) + \mathbb{P}((0, 0) \rightarrow (x, n), (0, 0) \not\Rightarrow (x, n)). \qquad (13.2)$$

In the last term, there must be an occupied pivotal bond for the connection, and hence a first such bond. Given a bond $b = \{(u, n), (v, n + 1)\}$, let $\bar{b} = (v, n + 1)$ be the "top" of b, and $\underline{b} = (u, n)$ be the "bottom" of b. Partitioning according to the first pivotal bond gives

$$\mathbb{P}((0, 0) \rightarrow (x, n), (0, 0) \not\Rightarrow (x, n))$$
$$= \sum_{b_1} \mathbb{P}((0, 0) \Rightarrow \underline{b}_1 \rightarrow \bar{b}_1 \rightarrow (x, n), (x, n) \notin \tilde{C}^{b_1}(0, 0)), \qquad (13.3)$$

where the condition $(x, n) \notin \tilde{C}^{b_1}(0, 0)$ ensures that b_1 is pivotal. We use inclusion-exclusion for this condition, and then the Markov property, to obtain

$$\mathbb{P}((0, 0) \Rightarrow \underline{b}_1 \rightarrow \bar{b}_1 \rightarrow (x, n), (x, n) \notin \tilde{C}^{b_1}(0, 0))$$
$$= \mathbb{P}((0, 0) \Rightarrow \underline{b}_1 \rightarrow \bar{b}_1 \rightarrow (x, n))$$
$$\quad - \mathbb{P}((0, 0) \Rightarrow \underline{b}_1 \rightarrow \bar{b}_1 \rightarrow (x, n), (x, n) \in \tilde{C}^{b_1}(0, 0))$$
$$= \mathbb{P}((0, 0) \Rightarrow \underline{b}_1) p \mathbb{P}(\bar{b}_1 \rightarrow (x, n))$$
$$\quad - \mathbb{P}((0, 0) \Rightarrow \underline{b}_1 \rightarrow \bar{b}_1 \rightarrow (x, n), (x, n) \in \tilde{C}^{b_1}(0, 0)). \qquad (13.4)$$

To rewrite (13.4), we define

$$\pi_m^{(0)}(x) = \mathbb{P}_p((0, 0) \Rightarrow (x, m)) - \delta_{0,x} \delta_{0,m}, \qquad (13.5)$$

and note that $\pi_0^{(0)}(x) = \pi_1^{(0)}(x) = 0$ by definition. For $n \geq 1$, the combination of (13.2)–(13.4) can be written as

$$\tau_n(x) = \pi_n^{(0)}(x) + (p|\Omega|D * \tau_{n-1})(x) + \sum_{m=2}^{n-1} (\pi_m^{(0)} * p|\Omega|D * \tau_{n-1-m})(x)$$
$$\quad - \sum_{b_1} \mathbb{P}((0, 0) \Rightarrow \underline{b}_1 \rightarrow \bar{b}_1 \rightarrow (x, n), (x, n) \in \tilde{C}^{b_1}(0, 0)). \qquad (13.6)$$

Note that if the last line of (13.6) is ignored, this is (13.1) with π replaced by $\pi^{(0)}$.

The expansion continues with the last term of (13.6). To understand this term, the concept of *backbone* is useful. Given a configuration in which $\bar{b}_1$ is connected to (x, n), the backbone is defined to consist of the vertices on occupied paths from $\bar{b}_1$ to (x, n) (including $\bar{b}_1$ and (x, n)). The event $\{(x, n) \in \tilde{C}^{b_1}(0, 0)\}$ is equivalent to the existence of a vertex $(y, m) \in \tilde{C}^{b_1}(0, 0)$ in the backbone. An important event is that all such vertices occur *after* any pivotal bonds for $\bar{b}_1 \to (x, n)$. This motivates the following definition.

Given a bond b, a vertex (x, n), and a random or deterministic set A of vertices, let $E(b, (x, n); A)$ denote the event that:

- b is occupied, and
- there exists a vertex $(y, m) \in A$ such that $\bar{b} \to (y, m) \to (x, n)$, and
- there is no pivotal bond b' for the connection from $\bar{b}$ to (x, n) such that the backbone joining $\bar{b}$ to $\underline{b}'$ contains a vertex of A.

This event plays a key role in the development of the expansion. We write $\bar{b} < \bar{b}'$ to mean that the temporal component of $\bar{b}$ is less than that of $\bar{b}'$, and, similarly, we write $\bar{b} < n$ when the temporal component of $\bar{b}$ is less than n.

Exercise 13.1. Let $w \in \mathbb{Z}^d$, and suppose that $l < \bar{b} \le n$. Show that the above event is related to the event of (10.11) by

$$E(b, (x, n); \tilde{C}^b(w, l)) = \{b \text{ is occupied}\} \cap E'(\bar{b}, (x, n); \tilde{C}^b(w, l)), \qquad (13.7)$$

where here the event $\tilde{C}^b(w, l)$ belongs to the same probability space as the events E and E' (i.e., there is a single percolation model and not multiple models as in Sect. 10.1).

Lemma 13.2. *Let b be a bond, let F be an event that depends only on bonds (u, j) with $j \le \underline{b}$, let (v, k) be a vertex with $k \le \underline{b}$, let (x, n) be a vertex with $\bar{b} \le n$, and, given any vertex (y, m) with $\bar{b} \le m$, let $F_{(y,m)} = F \cap E(b, (y, m); \tilde{C}^b(v, k))$. Then*

$$\mathbb{P}\big(F \cap \{\underline{b} \to \bar{b} \to (x, n)\} \cap \{(x, n) \in \tilde{C}^b(v, k)\}\big)$$
$$= \mathbb{P}(F_{(x,n)}) + \sum_{b'} \mathbb{P}(F_{\underline{b}'}) p \mathbb{P}(\bar{b}' \to (x, n))$$
$$- \sum_{b'} \mathbb{P}\big(F_{\underline{b}'} \cap \{\underline{b}' \to \bar{b}' \to (x, n)\} \cap \{(x, n) \in \tilde{C}^{b'}(\bar{b})\}\big). \qquad (13.8)$$

Proof. We partition the event on the left hand side according to the *first* pivotal bond b' for $\bar{b} \to (x, n)$, if there is one, such that the event $E(\bar{b}, \underline{b}'; \tilde{C}^b(v, k))$ occurs. If there is no such pivotal bond, then the event $E(\bar{b}, (x, n); \tilde{C}^b(v, k))$

occurs, producing the first term on the right hand side of (13.8). The remaining contribution is

$$\sum_{b'} \mathbb{P}\big(F_{\underline{b}'} \cap \{\underline{b}' \to \bar{b}' \to (x,n)\} \cap \{(x,n) \notin \tilde{C}^{b'}(\bar{b})\}\big). \qquad (13.9)$$

After using inclusion-exclusion and then the Markov property on the last event in (13.9), this gives (13.8). ∎

Note that the summand in the last term on the right hand side of (13.8) is of the same form as the left hand side, so the identity can be iterated. The iteration begins with the last term on the right hand side of (13.4), whose summand is equal to the left hand side of (13.8) with $F = \{(0,0) \Rightarrow \underline{b}_1\}$. To record the result of the iteration, we make the following definitions. For $N \geq 1$, let

$$B_N(n) = \{\vec{b} = (b_1, \ldots, b_N) : 0 < \bar{b}_1 < \cdots < \bar{b}_N \leq n\} \qquad (13.10)$$

denote the ordered vectors of N bonds, up to time n. Given $\vec{b} \in B_N(n)$, we define $\bar{b}_0 = (0,0)$, $\underline{b}_{N+1} = (x,n)$. For $N \geq 1$, we define

$$\pi_m^{(N)}(x) = \sum_{\vec{b} \in B_N(m)} \mathbb{P}_p\Big[\{(0,0) \Rightarrow \underline{b}_1\} \cap \bigcap_{i=1}^{N} E(\bar{b}_i, \underline{b}_{i+1}; \tilde{C}^{b_i}(\bar{b}_{i-1}))\Big]. \qquad (13.11)$$

Also, we define

$$\pi_m(x) = \sum_{N=0}^{\infty} (-1)^N \pi_m^{(N)}(x), \qquad (13.12)$$

where the apparently infinite sum is actually a finite sum, since the sum in (13.11) is empty if $N > m$, in which case $\pi_m^{(N)}(x) = 0$.

The last term of (13.6) is then computed iteratively using Lemma 13.2, to produce the expansion (13.1). The iteration eventually terminates because the remainder term in (13.8) will vanish after the number of iterations exceeds n.

13.2 Laces and Resummation

In this section, we derive the expansion of Nguyen and Yang [166]. This expansion has been applied also in [121, 167]. The method is closely related to the lace expansion for lattice trees and lattice animals, and uses the notion of connected graph and lace from Sect. 8.1.1.

For $t = 0, 1, 2, \ldots$, we define $W_{n,t}(x)$ to be the event that $(0,0) \to (x,n)$ with exactly t occupied pivotal bonds for the connection, and let

$$\tau_{n,t}(x) = \mathbb{P}_p(W_{n,t}(x)). \qquad (13.13)$$

By definition, $\tau_n(x) = \sum_{t=0}^{\infty} \tau_{n,t}(x)$. We will rewrite $\tau_{n,t}(x)$ in terms of a repulsive interaction between the sausages in the string of sausages representing the connection $(0,0) \to (x,n)$ (see Fig. 10.1). When $W_{n,t}(x)$ occurs, there are exactly $t+1$ sausages.

Recalling the definition of $B_t(n)$ in (13.10), for $t \geq 1$ and $\vec{b} \in B_t(n)$ we define

$$T(\vec{b},(x,n)) = \bigcap_{i=1}^{t}\{b_i \text{ occupied}\} \bigcap_{i=0}^{t}\{\bar{b}_i \Rightarrow \underline{b}_{i+1}\}. \qquad (13.14)$$

Note that if $T(\vec{b},(x,n))$ occurs, then the only possible candidates for occupied pivotal bonds for the event $(0,0) \to (x,n)$ are the elements of $\vec{b}$ (but these candidates need not be pivotal). We define the random variables

$$K[i,j] = \prod_{i \leq s < t \leq j} (1+U_{st}) \quad \text{with} \quad U_{ij} = -I[\bar{b}_i \Rightarrow \underline{b}_{j+1}]. \qquad (13.15)$$

The product in (13.15) is 0 or 1. The event that both $T(\vec{b},(x,n))$ occurs and $K[0,t] = 1$ is the event that the occupied pivotal bonds for $(0,0) \to (x,n)$ are precisely the elements of $\vec{b}$. Therefore, for $t \geq 1$,

$$\tau_{n,t}(x) = \sum_{\vec{b} \in B_t(n)} \mathbb{E}_p[I[T(\vec{b},(x,n))]K[0,t]]. \qquad (13.16)$$

Using the terminology of graphs and connected graphs defined in Sect. 8.1.1, we have

$$K[a,b] = \sum_{\Gamma \in \mathcal{B}[a,b]} \prod_{ij \in \Gamma} U_{ij}, \qquad (13.17)$$

with $K[a,a] = 1$. Let

$$J[a,b] = \sum_{\Gamma \in \mathcal{G}[a,b]} \prod_{ij \in \Gamma} U_{ij}. \qquad (13.18)$$

These are related by (8.6), which states that for $t \geq 1$,

$$K[0,t] = K[1,t] + \sum_{s=1}^{t-1} J[0,s]K[s+1,t] + J[0,t], \qquad (13.19)$$

where the empty sum is zero if $t = 1$. Let

$$\pi_{m,0}(x) = \mathbb{P}_p((0,0) \Rightarrow (x,m)) - \delta_{0,x}\delta_{0,m}, \qquad (13.20)$$

$$\pi_{m,s}(x) = \sum_{\vec{b} \in B_s(m)} \mathbb{E}_p[I[T(\vec{b},(x,m))]J[0,s]] \qquad (s \geq 1). \qquad (13.21)$$

It can be seen from the above definitions that $\pi_{m,0}(x) = \tau_{m,0}(x)$ if $m > 0$, and that $\pi_{m,s}(x) = 0$ whenever $m = 0, 1$ or $s > m$. Let

$$\pi_m(x) = \sum_{s=0}^{\infty} \pi_{m,s}(x) = \sum_{s=0}^{m} \pi_{m,s}(x). \qquad (13.22)$$

Substitution of (13.19) into (13.16), followed by application of the Markov property, then gives the recursion formula

$$\tau_{n,t}(x) = (p|\Omega|D * \tau_{n-1,t-1})(x) + \sum_{m=2}^{n-1}\sum_{s=0}^{t-1}(\pi_{m,s} * p|\Omega|D * \tau_{n-1-m,t-1-s})(x)$$
$$+ \pi_{n,t}(x), \tag{13.23}$$

valid for all $n, t \geq 1$. Summation of (13.23) over $t \geq 1$, and using $\pi_{n,0}(x) = \tau_{n,0}(x)$ for $n \geq 1$, then gives

$$\tau_n(x) = (p|\Omega|D * \tau_{n-1})(x) + \sum_{m=2}^{n-1}(\pi_m * p|\Omega|D * \tau_{n-1-m})(x) + \pi_n(x) \tag{13.24}$$

for $n \geq 1$. The identity (13.24) is identical to (13.1).

To obtain a useful representation for $\pi_m(x)$, we rewrite $\pi_{m,s}(x)$ in terms of laces. Insertion of (8.5) into (13.21) leads to

$$\pi_{m,s}(x) = \sum_{N=0}^{\infty}(-1)^N \pi_{m,s}^{(N)}(x), \tag{13.25}$$

with $\pi_{m,s}^{(0)}(x) = \pi_{m,0}(x)\delta_{0,s}$, and, for $N \geq 1$,

$$\pi_{m,s}^{(N)}(x) = \sum_{\vec{b} \in B_s(m)} \mathbb{E}_p\Big[I[T(\vec{b}, (x, m))]$$
$$\times \sum_{L \in \mathcal{L}^{(N)}[0,s]}\prod_{ij \in L}(-U_{ij})\prod_{i'j' \in \mathcal{C}(L)}(1 + U_{i'j'})\Big] \tag{13.26}$$

(the right hand side is interpreted as 0 when $s = 0$). The product $\prod_{ij \in L}(-U_{ij})$ is either 0 or 1, so $\pi_{m,s}^{(N)}(x) \geq 0$. The above gives the decomposition

$$\pi_m(x) = \sum_{N=0}^{\infty}(-1)^N \pi_m^{(N)}(x) \quad \text{with} \quad \pi_m^{(N)}(x) = \sum_{s=0}^{\infty}\pi_{m,s}^{(N)}(x). \tag{13.27}$$

With some effort, it can be seen that $\pi_m^{(N)}$ of (13.26)–(13.27) is the same as $\pi_m^{(N)}$ of (13.11).

13.3 Application of the Percolation Expansion

Recall that the expansion

$$\tau(0, y) = \delta_{0,y} + (p|\Omega|D * \tau)(0, y) + (\Pi_M * p|\Omega|D * \tau)(0, y) + \Pi_M(0, y) + R_M(0, y), \tag{13.28}$$

of (13.28), with Π_M and R_M given by (10.28)–(10.29), applies to an arbitrary regular graph. In the derivation of (13.28), bonds were not directed. However, the modifications due to orientation are simply notational, as we now explain.

For oriented percolation on $\mathbb{Z}^d \times \mathbb{Z}_+$, the variables 0 and y in (13.28) now correspond to space-time vertices $(0,0)$ and (x,n) (say). The convolution in (13.28) is a convolution on $\mathbb{Z}^d \times \mathbb{Z}_+$, and thus involves both spatial and temporal convolution.

The left hand side of (13.28) is simply the two-point function $\tau_n(x)$. We assume that $n \geq 1$, so the Kronecker delta on the right hand side of (13.28) is zero. In an abuse of notation, the function D in (13.28), which is a function on pairs $((u,j),(v,j+1))$, will be also written as $D(v-u)$. Thus, noting that the factor D consumes exactly one unit of time, the first convolution term in (13.28) can be written as the spatial convolution

$$p|\Omega|(D * \tau_{n-1})(x). \tag{13.29}$$

All connections in oriented percolation between distinct vertices involve a positive time difference between those vertices. In the definition of the remainder term $R_M((0,0),(x,n))$ in (10.29), there is a sum over oriented bonds $((u_i,m),(v_i,m+1))$, and the time associated with v_i cannot be more than the time associated with u_{i+1}, since events in (10.29) require connections from the former to the latter. The overall summation in (10.29) is therefore empty when $M > n$. The same is true for $\Pi^{(N)}$ for $N > m$, and we may simply take $M = \infty$ in (13.28), and drop the remainder term. Writing $\pi = \Pi_{M=\infty}$, we then have

$$(\Pi_M * p|\Omega|D * \tau)(0,y) + \Pi_M(0,y) = \sum_{m=2}^{\infty} (\pi_m * p|\Omega|D * \tau_{n-m-1})(x) + \pi_n(x) \tag{13.30}$$

(convolutions on the left are space-time and on the right space only). Thus (13.28) becomes (13.1).

The representations for $\Pi^{(N)}$ and $\pi^{(N)}$ are quite different and there is no apparent reason to expect them to be equal, even though their alternating sums must agree.

13.4 The Upper Critical Dimension

It is interesting to reinterpret for oriented percolation the explanation of the upper critical dimension given at the end of Sect. 10.1 for ordinary percolation. For oriented percolation, the approximation in which $\tau^{\tilde{C}^{\{u,v\}}(0)}(v,x)$ is replaced by by $\tau(v,x)$ is a good approximation when the backbone joining v and x typically does not have a space-time intersection with the cluster $\tilde{C}^{\{u,v\}}(0)$. Assuming Gaussian scaling limits, the backbone corresponds to a Brownian path, whereas the cluster $\tilde{C}^{\{u,v\}}(0)$ corresponds to a super-Brownian cluster (see Sect. 17.3). The upper critical dimension can then be understood as the dimension above which the *graphs* of Brownian motion and super-Brownian motion do not intersect.

Intersection of the graphs implies a collision of the two processes at the same time. It is known that $d = 4$ is critical for such a collision [19]. This can be understood heuristically by the following very rough argument. We first assume that since both processes are moving randomly it is reasonable to treat one of them as being stationary (this is a leap of faith). Regarding the super-Brownian motion as stationary, its support at fixed time is 2-dimensional. The Brownian path, which is two-dimensional, will generically not hit this support in dimensions greater than $4 = 2 + 2$. Alternately, if we regard the Brownian motion as being fixed, then its support is a point, hence 0-dimensional. The 4-dimensional range of super-Brownian motion will generically not hit this point in dimensions above $4 = 4 + 0$. Either way, this points to 4 as the upper critical dimension.

13.5 Diagrams for Oriented Percolation

If the representation for π_m of Sect. 13.3 is used, then π_m is bounded by diagrams precisely as in Sect. 10.2, with the understanding that now the two-point function is that of oriented percolation rather than ordinary percolation. These oriented percolation diagrams can then be estimated using the method of Sect. 10.3, for example, taking into account the orientation. There are different possibilities for how to proceed with this, and detailed estimates can be found in [119, 121, 166, 179].

It is worth noting that the representations for π_m of Sect. 13.1 and 13.2 can be estimated by diagrams that, although similar, are simpler than the ordinary percolation diagrams of Sect. 10.2. For example, consider $\pi_m^{(1)}(x)$ of (13.11). It can be concluded from the definition of the event $E((\bar{b}_1, (x, m); \tilde{C}^{b_1}(0))$ that the event

$$\{(0,0) \Rightarrow \underline{b}_1\} \cap E((\bar{b}_1, (x, m); \tilde{C}^{b_1}(0)) \qquad (13.31)$$

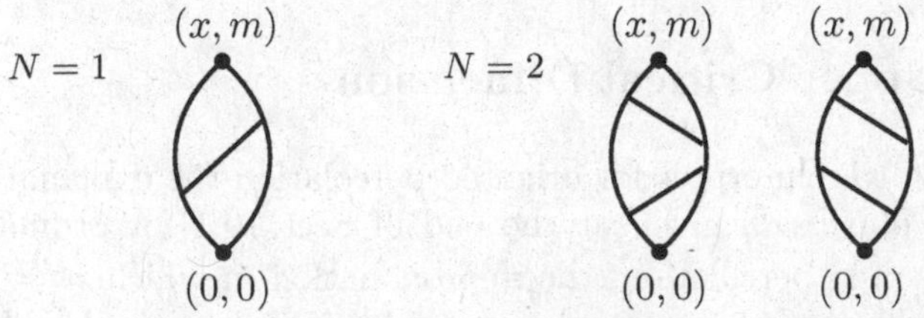

Fig. 13.1. Diagrams bounding $\pi_m^{(1)}(x)$ and $\pi_m^{(2)}(x)$.

is a subset of the event that there exists a vertex (z, l) and such that the following connections occur disjointly: $(0, 0) \to (z, l)$, $(z, l) \to \underline{b}_1$, $(0, 0) \to \underline{b}_1$, $\underline{b}_1 \to (x, n)$, and $(z, l) \to (x, n)$. Therefore $\pi_m^{(1)}(x)$ is bounded above by the left diagram in Fig. 13.1. It can similarly be concluded that $\pi_m^{(2)}(x)$ is bounded by the diagrams shown in Fig. 13.1, and this can be extended in a straightforward manner to obtain diagrams bounding $\pi_m^{(N)}(x)$ for larger values of N.

14

The Contact Process

The contact process is a much-studied model of the spread of infection, first introduced in [107]. For an introduction, see the books [153, 154]. In this section, we survey results that have been obtained for the critical behaviour of the contact process above its upper critical dimension $d = 4$.

14.1 The Phase Transition

The contact process is a continuous-time Markov process with state space $\{0,1\}^{\mathbb{Z}^d}$, with $d \geq 1$. In particular, the state of the contact process is determined by a variable $\xi_x \in \{0,1\}$, for each vertex $x \in \mathbb{Z}^d$. When $\xi_x = 0$, then x is considered "healthy," and when $\xi_x = 1$, then x is considered "infected." Infected particles spontaneously become healthy at rate 1, and healthy particles become infected at a rate proportional to their number of infected neighbours, i.e., at a rate $\lambda \sum_{y \in \Omega} \xi_{x+y}$, where we will assume that Ω is given either by (1.1) or (1.2).

The contact process has the convenient graphical representation depicted in Fig. 14.1. In the graphical representation, to each vertex there is associated a Poisson process of rate 1 called the recovery process, and to each directed bond (x,y) (with $y - x \in \Omega$) there is associated a Poisson process of rate λ called the infection process. All the Poisson processes are independent. Recovery marks are placed on the times lines of each vertex according to the recovery process at that vertex. Infection arrows are drawn from x to y at the times of the infection process associated to (x,y). As time proceeds, an infected particle at x becomes healthy at the recovery marks (nothing happens to a healthy particle at these times), and an infection arrow (x,y) causes y to become infected if x is infected. Thus, starting with an infected particle, infection proceeds along directed paths from that particle which follow time lines in the direction of increasing time, which do not cross recovery marks, and which follow infection arrows in the direction of the arrow.

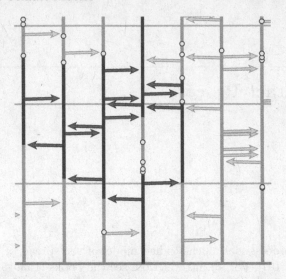

Fig. 14.1. The graphical representation of the contact process on $\mathbb{Z}^1$, with a single infected particle at time 0. Time increases in the upward direction.

We are interested in the situation in which at time zero there is a single infected particle at the origin and all other particles are healthy. The main question asks whether there is a positive probability that the infection survives for all time, or whether instead all particles are eventually healthy with probability one. The basic fact is that there is a critical value $\lambda_c \in (0, \infty)$ such that the infection dies out with probability one if $\lambda < \lambda_c$ but has a positive probability of survival if $\lambda > \lambda_c$. A theorem of Bezuidenhout and Grimmett [22] (see also [86] for the spread-out model) states that the infection dies out with probability one at the critical value $\lambda = \lambda_c$.

The phase transition can be studied in terms of several functions familiar from percolation. Let C_t denote the set of infected particles at time t. The two-point function is defined by

$$\tau_{\lambda,t}(x) = \mathbb{P}_\lambda(x \in C_t), \qquad (14.1)$$

for $x \in \mathbb{Z}^d$ and $t \geq 0$. The susceptibility is defined by

$$\chi(\lambda) = \int_0^\infty \mathrm{d}t\, \hat{\tau}_{\lambda,t}(0) = \mathbb{E}_\lambda \int_0^\infty \mathrm{d}t\, |C_t| = \mathbb{E}_\lambda \|C\|, \qquad (14.2)$$

where $\|C\| = \int_0^\infty \mathrm{d}t\, |C_t|$ is the total Lebesgue measure of all infection intervals on all time lines. The spread probability is defined by

$$\theta(\lambda) = \mathbb{P}_\lambda(C_t \neq \varnothing, \forall t \geq 0) = \mathbb{P}_\lambda(\|C\| = \infty). \qquad (14.3)$$

It is known [23] that the critical point has the dual characterization

$$\lambda_c = \sup\{\lambda : \chi(\lambda) < \infty\} = \inf\{\lambda : \theta(\lambda) > 0\}. \tag{14.4}$$

The magnetization is defined by

$$M_\lambda(h) = 1 - \mathbb{E}_\lambda \left(e^{-h\|C\|} \right). \tag{14.5}$$

Critical exponents are defined, as usual, by

$$\chi(\lambda) \sim c(\lambda_c - \lambda)^{-\gamma} \quad \text{as } \lambda \to \lambda_c^-, \tag{14.6}$$
$$M_{\lambda_c}(h) \sim ch^{1/\delta} \quad \text{as } h \to 0, \tag{14.7}$$
$$\theta(\lambda) \sim c(\lambda - \lambda_c)^\beta \quad \text{as } \lambda \to \lambda_c^+. \tag{14.8}$$

The critical exponents are predicted to be universal, and it is an important problem to prove that they exist and to calculate their values. The mean-field values, corresponding to the contact process on a regular tree or to branching random walk, are $\gamma = \beta = 1$ and $\delta = 2$. The mean-field bounds

$$\chi(\lambda) \geq c(\lambda_c - \lambda)^{-1}, \tag{14.9}$$
$$M_{\lambda_c}(h) \geq ch^{1/2}, \tag{14.10}$$
$$\theta(\lambda) \geq c(\lambda - \lambda_c) \tag{14.11}$$

are proved in [23] (for β and δ) and in [21] (for γ), for all $d \geq 1$. The hyperscaling inequalities proved in [180] imply that the upper critical dimension for the contact process is at least 4, in the sense that not all critical exponents can assume their mean-field values in dimensions less than 4 (assuming the exponents exist).

14.2 Approximation by Oriented Percolation

The contact process can be approximated by an oriented percolation model. This observation has been an important ingredient in the work of many authors, including [21, 23, 117, 119, 179].

To discretize time, we replace the time interval $[0, \infty)$ by $\epsilon \mathbb{Z}_+$, and define an oriented percolation model on $\mathbb{Z}^d \times \epsilon \mathbb{Z}_+$, as follows. Bonds have the form $((x, t), (y, t + \epsilon))$ with $t \in \epsilon Z_+$ and $y - x \in \{0\} \cup \Omega$. A bond is occupied with probability $1 - \epsilon$ if $x = y$, and with probability $\epsilon \lambda$ if $y - x \in \Omega$. Bonds with $x = y$ are called *temporal* and bonds with $x \neq y$ are called *spatial*. See Fig. 14.2.

Let $\mathbb{P}_\lambda^\epsilon$ denote the probability measure for the oriented percolation model on $\mathbb{Z}^d \times \epsilon \mathbb{Z}_+$. Then $\mathbb{P}_\lambda^\epsilon$ converges[1] weakly as $\epsilon \to 0^+$ to the contact process

---

[1] Fig. 14.1 is in fact a simulation of $\mathbb{P}_\lambda^\epsilon$ with small ϵ.

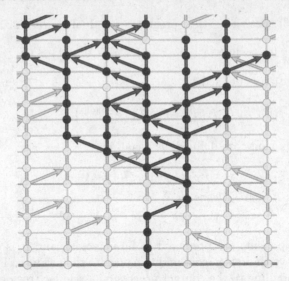

Fig. 14.2. Approximation of the contact process by oriented percolation on $\mathbb{Z}^1 \times \mathbb{Z}_+$.

measure $\mathbb{P}_\lambda$ (when formulated appropriately), and the critical value λ_c^ϵ of the discretized model converges to the critical value λ_c of the contact process (see [23, 179]).

This provides an avenue for analyzing the contact process by first analyzing an oriented percolation model on $\mathbb{Z}^d \times \epsilon\mathbb{Z}_+$ and then taking the limit $\epsilon \to 0^+$. In particular, the expansions of Chap. 13 for oriented percolation can be applied to the discretized model.

14.3 The Infrared Bound and the Triangle Condition

Let

$$\hat{\tau}_\lambda(k, \theta) = \sum_{x \in \mathbb{Z}^d} \int_0^\infty dt\, \tau_{\lambda,t}(x) e^{ik \cdot x} e^{i\theta t}, \qquad (14.12)$$

for $k \in [-\pi, \pi]^d$ and $\theta \in [-\pi, \pi]$. This is analogous to (12.15) with $z = e^{i\theta}$. The following infrared bound was proved in [179]. The proof uses discretization by oriented percolation, the oriented percolation expansion of Sect. 13.1, and a convergence proof based on the bootstrap Lemma 5.9.

Theorem 14.1. *There exist $d_0 \gg 4$ and $L_0(d) \gg 1$ such that for the nearest-neighbour contact process in dimensions $d \geq d_0$ or for the spread-out contact process in dimensions $d > 4$ with $L \geq L_0(d)$, there is a constant C (depending on d and L) such that the infrared bound*

$$|\hat{\tau}_\lambda(k,\theta)| \le \frac{C}{|k|^2 + |\theta|} \tag{14.13}$$

holds uniformly in $\lambda < \lambda_c$, $k \in [-\pi,\pi]^d$ *and* $\theta \in [-\pi,\pi]$.

The infrared bound implies the triangle condition for the contact process. To state the triangle condition, we define

$$\nabla_\lambda(R) = \sup_{x:|x| \ge R, t \ge 0} \sum_{y,z} \int_0^\infty ds_1 \int_0^\infty ds_2 \tau_{\lambda,s_1}(y) \tau_{\lambda,s_2-s_1}(z-y) \tau_{\lambda,s_2-t}(z-x). \tag{14.14}$$

The triangle condition is the statement that

$$\lim_{R \to \infty} \nabla_{\lambda_c}(R) = 0, \tag{14.15}$$

and it is proved in [21] that if the triangle condition holds then the exponents γ, β, δ exist and take their mean-field values, in the sense that there are upper bounds (with different constants) complementary to the lower bounds (14.9)–(14.11). As in Exercise 12.2, if $d > 4$ then the infrared bound (14.13) implies the triangle condition. Therefore, under the high-d hypotheses of Theorem 14.1, the critical exponents γ, β, δ exist (in the sense of $\simeq$) and take the mean-field values $\gamma = \beta = 1$ and $\delta = 2$.

For the spread-out model, this was extended in [119] to prove the following theorem. The proof makes use of the induction method of [120], adapted to incorporate uniformity in the discretization parameter ϵ.

Theorem 14.2. *Consider the spread-out model of the contact process. Let* $d > 4$, $\lambda = \lambda_c$ *and* $\delta \in (0, 1 \wedge \frac{d-4}{2})$. *There is an* $L_0 = L_0(d)$ *such that for* $L \ge L_0$ *there exist finite positive constants* v *and* A *(depending on* d *and* L) *and* C_1, C_2 *(depending only on* d) *such that the following statements hold as* $t \to \infty$:
(a)

$$\hat{\tau}_{\lambda_c,t}(k/\sqrt{vt}) = Ae^{-|k|^2/2d}\left[1 + O(|k|^2 t^{-\delta}) + O(t^{-(d-4)/2})\right], \tag{14.16}$$

(b)

$$\frac{1}{\hat{\tau}_{\lambda_c,t}(0)} \sum_{x \in \mathbb{Z}^d} |x|^2 \tau_{\lambda_c,t}(x) = vt\left[1 + O(t^{-\delta})\right], \tag{14.17}$$

(c)

$$C_1 L^{-d} t^{-d/2} \le \sup_{x \in \mathbb{Z}^d} \tau_{\lambda_c,t}(x) \le e^{-t} + C_2 L^{-d} t^{-d/2}, \tag{14.18}$$

with the error estimate in (a) uniform in $k \in \mathbb{R}^d$ *with* $|k|^2/\log t$ *sufficiently small.*

An extension of Theorem 14.2 to r-point functions for all $r \geq 2$, due to [117], is discussed in Sect. 17.4.

As is typical when the induction method of [120] is applied, the results of [119] also include the statements that $A = 1 + O(L^{-d})$, that $v = \sigma^2[1 + O(L^{-d})]$ where $\sigma^2 = \sum_x |x|^2 D(x)$ is the variance of D, and that the critical value obeys $\lambda_c = |\Omega|^{-1}[1 + O(L^{-d})]$. The latter is improved in [118][2] to

$$\lambda_c = \frac{1}{|\Omega|}\left[1 + \frac{1}{L^d}\sum_{n=2}^{\infty} U^{*n}(0) + O(L^{-(d+1)})\right], \tag{14.19}$$

where U is the probability density function of a uniform random variable on $[-1, 1]^d$ and U^{*n} denotes its n-fold convolution. This improves a result of [68], where (14.19) was obtained with error term $o(|\Omega|^{-1}) = o(L^{-d})$ by completely different methods. On the other hand, the results of [68] gives estimates for λ_c also in dimensions $2 \leq d \leq 4$.

A related model is studied in [68, 119]. In this model, the infection range L is assumed to scale as

$$L = L_T = L_1 T^b, \tag{14.20}$$

where b is a fixed parameter. This model allows for an analysis in dimensions $1 \leq d \leq 4$, provided b is large enough. The result is consistent with the general philosophy that the upper critical dimension is lowered when the range of interaction is suitably increased. The following theorem is proved in [119].

Theorem 14.3. *Let* $1 \leq d \leq 4$, $b > \frac{4-d}{2d}$, *and fix* $\delta \in (0, 1 \wedge bd + \frac{d-4}{2})$. *Then there exist* $\alpha > 0$, $L_0 \gg 1$, *and a critical value* λ_T, *such that for* $\lambda = \lambda_T$ *and* $L_1 \geq L_0$ *there exist* C_1, C_2 *(depending only on* d*) such that for all* $0 < t \leq \log T$, *as* $T \to \infty$,
(a)

$$\hat{\tau}_{\lambda_T, Tt}(k/\sqrt{v_T Tt}) = \mathrm{e}^{-|k|^2/2d}\left[1 + O(|k|^2(1+Tt)^{-\delta}) + O(T^{-\alpha})\right], \tag{14.21}$$

(b)

$$\frac{1}{\hat{\tau}_{\lambda_T, Tt}(0)}\sum_x |x|^2 \tau_{\lambda_T, Tt}(x) = v_T Tt\left[1 + O((1+Tt)^{-\delta}) + O(T^{-\alpha})\right], \tag{14.22}$$

(c)

$$C_1 L_T^{-d}(1+Tt)^{-d/2} \leq \sup_{x \in \mathbb{Z}^d} \tau_{\lambda_T, Tt}(x) \leq \mathrm{e}^{-Tt} + C_2 L_T^{-d}(1+Tt)^{-d/2}, \tag{14.23}$$

where v_T *is the variance of the range-*L_T D*. The error estimate in (14.16) is uniform in* $k \in \mathbb{R}^d$ *with* $|k|^2(\log(2+Tt))^{-1}$ *sufficiently small.*

---

[2] Note that p_c in [118] is equal to our $\lambda_c|\Omega|$.

The assumption on b in Theorem 14.3 is

$$b > \frac{4-d}{2d} = \begin{cases} 3/2 \text{ if } d = 1, \\ 1/2 \text{ if } d = 2, \\ 1/6 \text{ if } d = 3, \\ 0 \quad \text{ if } d = 4. \end{cases} \tag{14.24}$$

Related results were obtained in [68] assuming $b = 1$ in dimensions $d \geq 3$. On the other hand, in [68], for $d = 2$ the weaker assumption $L_T = (T \log T)^{1/2}$ is required, whereas Theorem 14.3 requires $b > \frac{1}{2}$ when $d = 2$.

To get some idea how the condition $b > \frac{4-d}{2d}$ arises, consider the following. A quantity related to the critical triangle diagram (14.14) is

$$\sum_{y,z} \int_0^{T \log T} ds_2 \int_0^{s_2} ds_1 \tau_{s_1}(y) \tau_{s_2 - s_1}(z - y) \tau_{s_2}(z), \tag{14.25}$$

where we consider $\lambda = \lambda_T$, and where we have taken $x = 0$ and cut off the integral at the maximum value $T \log T$ allowed for Tt in Theorem 14.3. The estimates of Theorem 14.3 give good bounds on this quantity when $b > \frac{4-d}{2d}$. In fact, using these estimates (with $k = 0$ in (14.21)), (14.25) is at most a multiple of

$$\int_0^{T \log T} ds_2 \int_0^{s_2} ds_1 \left(e^{-s_2} + \frac{1}{L_T^d (1 + s_2)^{d/2}} \right) \leq O(1 + T^{-bd}(T \log T)^{2 - d/2}), \tag{14.26}$$

and the condition $b > \frac{4-d}{2d}$ ensures that the right hand side is finite.

Remark 14.4. Theorems 14.2 and 14.3 are proved by obtaining versions of the theorems for the discretized model, with the estimates in the theorems uniform in $\epsilon \in (0,1]$. In particular, the results hold when $\epsilon = 1$, which proves versions of the theorems, including Theorem 14.3, for the spread-out oriented percolation model with the usual unit time discretization.

14.4 Expansion for the Contact Process

In Chap. 13, we focussed attention on the nearest-neighbour and spread-out oriented percolation models, but the analysis holds more generally. In particular, the expansion (13.1) also applies to the oriented percolation model obtained by discretizing the contact process, when suitably interpreted. Two modifications are required to make (13.1) apply to the discretized contact process. The first is replacement of $p|\Omega|D(x)$ in the first two terms of the right hand side by

$$p_\lambda^\epsilon(x) = \lambda \epsilon |\Omega| D(x) + (1 - \epsilon)\delta_{x,0}, \tag{14.27}$$

where the first and second terms on the right hand side correspond to the spatial and temporal bonds, respectively. The second modification is to interpret the definition of π_m in terms of the oriented percolation model with both spatial and temporal bonds.

The expansion for the discretized contact process can then be written down immediately from (13.1), as

$$\tau_{n\epsilon}^{\epsilon}(x) = (p^{\epsilon} * \tau_{(n-1)\epsilon}^{\epsilon})(x) + \sum_{m=2}^{n-1}(\pi_{m\epsilon}^{\epsilon} * p^{\epsilon} * \tau_{(n-1-m)\epsilon}^{\epsilon})(x) + \pi_{n\epsilon}^{\epsilon}(x). \quad (14.28)$$

Here, and in the following, we omit subscripts λ from $\tau_{n\epsilon}^{\epsilon}(x)$, $p^{\epsilon}(x)$ and $\pi_{m\epsilon}^{\epsilon}(x)$. If we consider momentarily the much simpler problem for which $\pi_{m\epsilon}^{\epsilon}$ is identically zero, then (14.28) becomes

$$\tau_{n\epsilon}^{\epsilon}(x) = \lambda\epsilon|\Omega|(D * \tau_{(n-1)\epsilon}^{\epsilon})(x) + (1-\epsilon)\tau_{(n-1)\epsilon}^{\epsilon}(x). \quad (14.29)$$

In terms of the Fourier transform, this can be rewritten as

$$\frac{\hat{\tau}_{n\epsilon}^{\epsilon}(k) - \hat{\tau}_{(n-1)\epsilon}^{\epsilon}(k)}{\epsilon} = \lambda|\Omega|\hat{D}(k)\hat{\tau}_{(n-1)\epsilon}^{\epsilon}(k) - \hat{\tau}_{(n-1)\epsilon}^{\epsilon}(k). \quad (14.30)$$

In the limit $\epsilon \to 0^{+}$, where we also take $n\epsilon \to t$, this formally becomes the differential equation

$$\frac{\mathrm{d}}{\mathrm{d}t}\hat{\tau}_t(k) = -[1 - \lambda|\Omega|\hat{D}(k)]\hat{\tau}_t(k). \quad (14.31)$$

With the initial condition $\hat{\tau}_0(k) = 1$, (14.31) has solution

$$\hat{\tau}_t(k) = \mathrm{e}^{-[1-\lambda|\Omega|\hat{D}(k)]t}. \quad (14.32)$$

For $\lambda = 1/|\Omega|$, this is the Fourier transform of the transition probability for continuous-time random walk which jumps at rate 1.

If we now restore the omitted terms involving π^{ϵ} in (14.28), then the right hand side of (14.30) contains the additional terms

$$\epsilon\sum_{m=2}^{n-1}\frac{1}{\epsilon^2}\hat{\pi}_{m\epsilon}^{\epsilon}(k)\hat{p}^{\epsilon}(k)\hat{\tau}_{(n-m-1)\epsilon}^{\epsilon}(k) + \frac{1}{\epsilon}\hat{\pi}_{n\epsilon}^{\epsilon}(k). \quad (14.33)$$

The factor ϵ has been placed in front of the summation to produce a Riemann sum. Since $\lim_{\epsilon\to 0}\hat{p}_{\lambda}^{\epsilon}(k) = 1$, the limit of (14.33) apparently takes the form

$$\int_0^t \hat{\pi}_s(k)\hat{\tau}_{t-s}(k)\mathrm{d}s, \quad (14.34)$$

with, for $m\epsilon \to s$ as $\epsilon \to 0$,

$$\hat{\pi}_s(k) = \lim_{\epsilon \to 0} \frac{1}{\epsilon^2}\hat{\pi}^\epsilon_{m\epsilon}(k) \tag{14.35}$$

(note that the last term of (14.33) vanishes in the limit if (14.35) holds.) Thus, provided it is possible to make sense of the limit (14.35), we expect that the contact process two-point function obeys the integro-differential equation

$$\frac{\mathrm{d}}{\mathrm{d}t}\hat{\tau}_t(k) = -[1 - \lambda|\Omega|\hat{D}(k)]\hat{\tau}_t(k) + \int_0^t \hat{\pi}_s(k)\hat{\tau}_{t-s}(k)\mathrm{d}s. \tag{14.36}$$

A proof that this all works under the hypotheses of Theorem 14.2 or 14.3 is given in [119].

Equation (14.36) is a continuous-time version of the lace expansion, and it would be interesting to analyze it directly. However, such an analysis has not been carried out. Instead, in [119,179], properties of the solution to (14.28) are analyzed directly, and estimates uniform in ϵ are obtained. The limit $\epsilon \to 0$ can then be taken to prove Theorems 14.1–14.3. A crucial step in the analysis is to prove that $\hat{\pi}^\epsilon_{m\epsilon}(k)$ is $O(\epsilon^2)$, consistent with (14.35). Next, we discuss the main idea that permits this.

14.5 Diagrams for the Contact Process

Now we describe the new observation needed to prove that $\hat{\pi}^\epsilon_s(k)$ is $O(\epsilon^2)$, consistent with (14.35), where we write $s = m\epsilon$. The diagrams for the discretized contact process are the same as the oriented percolation diagrams discussed in Sect. 13.5, with the understanding that temporal bonds now also occur.

The key observation already arises in the $N = 0$ term, which is (cf. (13.5))

$$(\hat{\pi}^\epsilon_s)^{(0)}(k) = \sum_{x \in \mathbb{Z}^d} \mathbb{P}^\epsilon_\lambda((0,0) \Rightarrow (x,s))e^{ik \cdot x} \tag{14.37}$$

for $s = m\epsilon > 0$. Given that there are two disjoint connections from $(0,0)$ to (x,s), and given that there is only one temporal bond leaving $(0,0)$ and one temporal bond entering (x,s), it must be the case that one of the connections uses a spatial bond to leave $(0,0)$ and one must use a spatial bond to enter (x,s). Thus there are two possibilities, depending on whether these two spatial bonds are part of the same connection or part of disjoint connections. This leads to the upper bound

$$\lambda^2|\Omega|^2\epsilon^2 \sum_{x \in \mathbb{Z}^d} \sum_{u \in \Omega} \sum_{x' \in x - \Omega} D(u)D(x - x')$$
$$\times \left[\tau^\epsilon_{s-\epsilon}(x - u)\tau^\epsilon_{s-\epsilon}(x') + \tau^\epsilon_s(x)\tau^\epsilon_{s-2\epsilon}(x' - u)\right] \tag{14.38}$$

for the absolute value of (14.37). If we now assume that the estimates (14.16) and (14.18) apply to the discretized model, and that $\lambda|\Omega| = O(1)$, then we can bound this above by $O(\epsilon^2 L^{-d} s^{-d/2})$. In more detail, for the second term, we use (14.18) to bound $\tau_s^\epsilon(x)$ by $O(L^{-d} s^{-d/2})$, and then bound the sums using (14.16). Here we have used the theorem whose method of proof we are discussing, but this can be handled inductively.

For the higher-order diagrams that bound $\hat{\pi}_s^\epsilon(k)$, each diagram vertex contributes a factor ϵ due to the fact that there is only one temporal bond entering and leaving any given vertex. For all vertices other than the origin and the top vertex, this factor ϵ combines with the summation over the temporal coordinate of the vertex to produce an overall contribution $O(1)$. Thus, in the end, the upper bound involves $O(\epsilon^2)$. This procedure is carried out in detail in [119] and [179] for different forms of the diagrams (space-time diagrams and generating function diagrams, respectively).

15

Branching Random Walk

Branching random walk serves as the mean-field model for many interacting models involving branching, including lattice trees and percolation. In Sect. 15.1, we consider a natural mean-field model of lattice trees, which turns out to be intimately related to branching random walk with Poisson offspring distribution. In Sect. 15.2 we compute several generating functions important for a particular model of branching random walk. These will be used to derive the scaling limit of the branching random walk model in Chap. 16. In Sect. 15.3, we define a model of weakly self-avoiding lattice trees in terms of branching random walk.

15.1 A Mean-Field Model

The mean-field model for the self-avoiding walk is simple random walk, which forgets about the self-avoidance interaction. It is natural to attempt to define a mean-field model of lattice trees by somehow forgetting about the mutual avoidance of the branches in a lattice tree. In this section, we define such a mean-field model, as in [61]. For simplicity, we consider only the nearest-neighbour model.

It is convenient to switch to a site activity (or "fugacity"), rather than a bond activity. This means that we weight vertices rather than bonds by z, so that the one-point function of (7.20) becomes now $g(z) = G_z^{(1)} = \sum_{T:T\ni 0} z^{|T|+1}$. If we remove the product containing the interaction in the two-point function (8.2), we obtain $\sum_{\omega\in\mathcal{W}(0,x)}(g(z))^{|\omega|+1}$. It is also useful to keep track of the length of the path ω by associating an activity ζ to each bond in this path. This prompts us to define the two-point function of the mean-field model by

$$F_{z,\zeta}(x) = \sum_{\omega\in\mathcal{W}(0,x)}(f(z))^{|\omega|+1}\left(\frac{\zeta}{2d}\right)^{|\omega|}, \qquad (15.1)$$

where the function $f(z)$, to be specified below, is the one-point function for the mean-field model, and where the factor $\frac{1}{2d}$ has been included as a convenient normalization. The sum in (15.1) is taken over simple random walks. Thus, $F_{z,\zeta}(x)$ can be written in terms of the simple random walk two-point function (1.17) as $F_{z,\zeta}(x) = f(z)C_{\zeta f(z)/2d}(0,x)$, and hence, by (1.18), its Fourier transform is

$$\hat{F}_{z,\zeta}(k) = \frac{f(z)}{1 - \zeta f(z)\hat{D}(k)}, \tag{15.2}$$

with $\hat{D}(k)$ given by (1.12). The random walk two-point function has critical value $\frac{1}{2d}$, corresponding to $\zeta f(z) = 1$. We will realize the latter with $\zeta = f(z) = 1$.

For the function $f(z)$, we require by analogy with (7.22) (taking into account the switch to site activity), that $f(z)$ satisfy the differential equation

$$\hat{F}_{z,1}(0) = z\frac{\mathrm{d}f(z)}{\mathrm{d}z}. \tag{15.3}$$

Combining (15.3) and (15.2) gives

$$\frac{f(z)}{1 - f(z)} = z\frac{\mathrm{d}f(z)}{\mathrm{d}z}. \tag{15.4}$$

Integrating the separable equation (15.4) over an interval $[z, z_0]$ gives

$$\frac{f(z)e^{-f(z)}}{f(z_0)e^{-f(z_0)}} = \frac{z}{z_0}. \tag{15.5}$$

The initial condition $z_0 = 1$ and $f(z_0) = 1$ is a choice of normalization and gives

$$f(z)e^{-f(z)} = ze^{-1}. \tag{15.6}$$

By (15.6), f can be written as $f(z) = -W(-ze^{-1})$, where W is the principal branch of the Lambert W function defined by $W(w)e^{W(w)} = w$ [56]. The latter is analytic on the w-plane with branch cut $(-\infty, -e^{-1}]$, corresponding to a branch cut $[1, \infty)$ for $f(z)$ (and gives $f(0) = 0$, as it should). This uniquely specifies $f(z)$, and hence $\hat{F}_{z,\zeta}(k)$. The *same* functions $f(z)$ and $\hat{F}_{z,\zeta}(k)$ will arise below in Theorem 15.2 for a model of branching random walk.

15.2 Branching Random Walk

In this section, we define a model of branching random walk in terms of embeddings of trees into $\mathbb{Z}^d$. The presentation is based on [30]. Some related ideas can be found in [36].

The trees are the family trees of the critical Galton–Watson branching process with Poisson offspring distribution. In more detail, we begin with a single individual having ξ offspring, where ξ is a Poisson random variable of mean 1, i.e., $\mathbb{P}(\xi = m) = (em!)^{-1}$. Each of the offspring then independently has offspring of its own, with the same critical Poisson distribution. To indicate when two trees are the same, we describe them in terms of *words*. The root is the word 0. The children of the root are the words $01, 02, \ldots 0\xi_0$. The children of 01 are the words $011, \ldots, 01\xi_{01}$, and so on. A tree is then uniquely represented by a set of words, and two trees are the same if and only if they are represented by the same set of words. A tree T consisting of exactly n individuals, with the ith individual having ξ_i offspring, has probability

$$\mathbb{P}(T) = \prod_{i \in T} \frac{1}{e\,\xi_i!} = e^{-n} \prod_{i \in T} \frac{1}{\xi_i!}. \tag{15.7}$$

The product in (15.7) is over the vertices of T.

We define an embedding φ of T into $\mathbb{Z}^d$ to be a mapping from the vertices of T into $\mathbb{Z}^d$, such that the root is mapped to the origin and adjacent vertices in the tree are mapped to nearest neighbours in $\mathbb{Z}^d$. There is no assumption that φ is injective, and different vertices of T can be mapped to the same vertex in $\mathbb{Z}^d$. Given a tree T having $|T|$ vertices, there are $(2d)^{|T|-1}$ possible embeddings φ of T. A branching random walk configuration is then a pair (T, φ), with associated probability

$$\mathbb{P}(T, \varphi) = \frac{1}{(2d)^{|T|-1}} \mathbb{P}(T). \tag{15.8}$$

Our aim in this section is to compute the r-point functions of this branching random walk model. These are generating functions for trees of fixed total number of vertices n, which visit a specified set of $r - 1$ vertices in a specified manner (the rth point is the origin, where the embedding is rooted).

We begin with the simplest case $r = 1$. For $z \in \mathbb{C}$ with $|z| \leq 1$, the one-point function is defined by

$$b_z^{(1)} = \sum_{(T, \varphi)} \mathbb{P}(T, \varphi) z^{|T|} = \sum_T \mathbb{P}(T) z^{|T|}. \tag{15.9}$$

The series on the right hand side of (15.9) is the generating function for the probability mass function for the total size of a critical Poisson tree. It converges for $|z| \leq 1$, with $b_1^{(1)} = 1$. For general z, $b_z^{(1)}$ is given in the following theorem.

We write $p_m = \mathbb{P}(\xi = m) = (em!)^{-1}$, and let

$$P(w) = \sum_{m=0}^{\infty} p_m w^m = e^{w-1} \tag{15.10}$$

denote the generating function for the critical Poisson distribution.

Theorem 15.1. *For $d \geq 1$, the one-point function is given by*

$$b_z^{(1)} = \sum_{n=1}^{\infty} \frac{n^{n-1}}{n!} e^{-n} z^n, \tag{15.11}$$

and obeys

$$b_z^{(1)} e^{-b_z^{(1)}} = z e^{-1}. \tag{15.12}$$

Proof. Conditioning on the number of offspring of the root gives

$$b_z^{(1)} = \sum_{m=0}^{\infty} p_m z \left(b_z^{(1)} \right)^m = z P(b_z^{(1)}) = z e^{b_z^{(1)} - 1}, \tag{15.13}$$

which implies (15.12). The Taylor expansion (15.11) then follows from Lagrange's inversion formula (see, e.g., [196, p.43]). ∎

Since $b_z^{(1)}$ is real for real z and $b_0^{(1)} = 0$, it follows from (15.12) that $b_z^{(1)}$ is identical to the function $f(z)$ of the mean-field model of Sect. 15.1. Theorem 15.1 rederives the standard result that for the critical Poisson branching process,

$$\mathbb{P}(|T| = n) = \frac{n^{n-1}}{n!} e^{-n}. \tag{15.14}$$

By Stirling's formula,

$$\mathbb{P}(|T| = n) \sim \frac{1}{\sqrt{2\pi}} \frac{1}{n^{3/2}}. \tag{15.15}$$

Comparing with (7.2) and (7.3), this is a statement that the critical exponent θ takes the value $\theta = \frac{5}{2}$ for branching random walk.

The two-point function is a generating function for critical Poisson branching random walk which starts at the origin, which has a family tree whose total size is dual to an activity z, and which visits the vertex x (possibly more than once) at a time dual to an activity ζ. The two-point function is defined for $z, \zeta \in \mathbb{C}$ with $|z| < 1$, $|\zeta| \leq 1$, and for $x \in \mathbb{Z}^d$, by

$$b_{z,\zeta}^{(2)}(x) = \sum_{(T,\varphi)} \mathbb{P}(T,\varphi) z^{|T|} \sum_{i \in T} I[\varphi(i) = x] \zeta^{|i|}, \tag{15.16}$$

where $|i|$ denotes the graph distance from i to the root of T. The series (15.16) clearly converges for $|z| < 1$, $|\zeta| \leq 1$, as does its sum over $x \in \mathbb{Z}^d$. The following theorem gives the Fourier transform of the two-point function, and shows that it is identical to the mean-field two-point function of Sect. 15.1.

Theorem 15.2. *For $d \geq 1$, $k \in [-\pi, \pi]^d$, $|z| < 1$, $|\zeta| \leq 1$,*

$$\hat{b}_{z,\zeta}^{(2)}(k) = \frac{b_z^{(1)}}{1 - \zeta b_z^{(1)} \hat{D}(k)}. \tag{15.17}$$

The denominator of the right hand side vanishes for $z = \zeta = 1$, $k = 0$, and in that case $\hat{b}_{1,1}^{(2)}(0) = \infty$.

Proof. The contribution to the right hand side of (15.16) arising when i is the root is simply $b_z^{(1)} \delta_{0,x}$. When i is not the root, we condition on the number of offspring of the root and on the location of the first step on the branch containing i, to obtain

$$b_{z,\zeta}^{(2)}(x) = b_z^{(1)} \delta_{0,x} + \sum_{m=1}^{\infty} p_m m \left(b_z^{(1)} \right)^{m-1} z (\zeta D * b_{z,\zeta}^{(2)})(x)$$

$$= b_z^{(1)} \delta_{0,x} + z P'(b_z^{(1)}) (\zeta D * b_{z,\zeta}^{(2)})(x). \qquad (15.18)$$

In the second term of the middle member of (15.18), the factor z is associated with the root, and the factor m corresponds to choosing which of the root's offspring is an ancestor of the vertex j. The Poisson generating function obeys $z P'(b_z^{(1)}) = z P(b_z^{(1)})$, and by (15.13), this equals $b_z^{(1)}$. Taking the Fourier transform of (15.18) converts the convolution into a product, and we can then solve for $\hat{b}_{z,\zeta}^{(2)}(k)$ to obtain

$$\hat{b}_{z,\zeta}^{(2)}(k) = \frac{b_z^{(1)}}{1 - \zeta b_z^{(1)} \hat{D}(k)}. \qquad (15.19)$$

Note that the denominator is zero when $z = \zeta = 1$ and $k = 0$.

Finally, we observe that the Fourier transform of (15.16) is given by

$$\hat{b}_{z,\zeta}^{(2)}(k) = \sum_{(T,\varphi)} \sum_{j \in T} \mathbb{P}(T,\varphi) z^{|T|} e^{ik \cdot \varphi(j)} \zeta^{|j|}. \qquad (15.20)$$

We conclude from this and (15.15) that

$$\hat{b}_{1,1}^{(2)}(0) = \sum_{(T,\varphi)} \sum_{j \in T} \mathbb{P}(T,\varphi) = \sum_{n=1}^{\infty} n \mathbb{P}(|T| = n) = \infty. \qquad (15.21)$$

∎

The two-point function given in Theorem 15.2 can be interpreted as the two-point function of simple random walk with an activity ζ associated to each step of the walk and an activity $b_z^{(1)}$ associated to each vertex. We may therefore regard a critical Poisson branching random walk configuration containing 0 and x as corresponding to a simple random walk path from 0 to x with a one-point function attached at each vertex along the way. This was the philosophy of the mean-field model of Sect. 15.1.

Next, we define the *r*-point functions for $r \geq 3$. Our definition keeps track of a substantial amount of information, and requires as preparation the following definitions of *shape*, *subshape*, *skeleton* and *compatibility*.

Shape: Shapes are certain rooted binary trees. For $r \geq 2$, we give a recursive definition of the set Σ_r of *r-shapes*, as follows. Each *r*-shape has $2r - 3$ edges,

$r - 2$ vertices of degree 3 (the *branch points*) and r vertices of degree 1 (the *leaves*) labelled $0, 1, \ldots, r - 1$. There is a unique 2-shape given by the tree consisting of vertex 0 joined by a single edge to vertex 1. We think of this shape as indicating that vertex 0 is an ancestor of vertex 1. There is a unique 3-shape, consisting of three vertices $0, 1, 2$ each joined by an edge to a fourth vertex. We think of this shape as indicating that 0 is an ancestor of both 1 and 2. In general, for $r \geq 3$, to each $(r-1)$-shape σ, we obtain $2r - 5$ r-shapes by choosing one of the $2r - 5$ edges of σ, adding a vertex on that edge together with a new edge that joins the added vertex to a new leaf $r - 1$. The resulting r-shapes represent the different ways in which an additional r^{th} particle can be added to the family tree of $r - 1$ particles represented by σ. Thus there is a unique shape for $r = 2$ and $r = 3$, and $(2r - 5)!!$ distinct shapes for $r \geq 4$, where we use the notation $(-1)!! = 1$, and, for $r \geq 3$,

$$(2r - 5)!! = \prod_{j=3}^{r} (2j - 5). \tag{15.22}$$

When r is clear from the context, we will refer to an r-shape simply as a shape. For notational convenience, we associate to each shape an arbitrary labelling of its $2r - 3$ edges, with labels $1, \ldots, 2r - 3$. This arbitrary choice of edge labels is fixed once and for all; see Fig. 15.1.

Subshape: A subshape of a shape $\sigma \in \Sigma_r$ is a tree obtained by contracting a subset of the edges of σ to a point. This can lead to multiply-labelled vertices, and contracted edges lose their labels. The subshapes for $r = 3$ are shown in Fig. 15.2. In general, there are 2^{2r-3} subshapes of a shape $\sigma \in \Sigma_r$. We denote subshapes by λ and write $\lambda \leq \sigma$ when λ is a subshape of σ. We denote the set of edge labels of λ by $e(\lambda)$.

Skeleton: We write $\bar{\imath} = (i_1, \ldots, i_{r-1})$ for a sequence of $r - 1$ vertices i_j (not necessarily distinct) in a tree T, and define the skeleton B of $(T, \bar{\imath})$ to be the subtree of T spanning $0, i_1, \ldots, i_{r-1}$. We will distinguish $r - 1$ and $2r - 3$ component vectors by using $\bar{\ }$ and $\vec{\ }$, respectively.

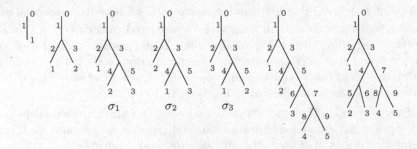

Fig. 15.1. The shapes for $r = 2, 3, 4$, and examples of the $7!! = 7 \cdot 5 \cdot 3 = 105$ shapes for $r = 6$. The shapes' edge labels are arbitrary but fixed.

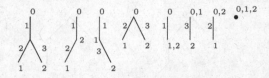

Fig. 15.2. The $2^3 = 8$ subshapes for $r = 3$.

Let β_B denote the tree obtained from B by ignoring vertices of degree 2 in B other than $0, i_1, \ldots, i_{r-1}$ (which may have degree 2 in B), and by assigning label j to vertex i_j for each j. This may lead to multiple labels at a vertex, as is the case for subshapes.

Given $\sigma \in \Sigma_r$ and a subshape $\lambda \le \sigma$, we say that β_B is isomorphic to λ if there is an edge preserving bijection from the set of all vertices of β_B to the set of all vertices of λ, which preserves the vertex labels of β_B and λ (including any multiple labels at vertices). Given such an isomorphism, the edge labels of λ induce labels on the edges of β_B and thus on the paths in T comprising the skeleton B.

Compatibility: Let $\sigma \in \Sigma_r$, $\vec{m} = (m_1, \ldots, m_{2r-3})$ for non-negative integers m_j, and $\vec{y} = (y_1, \ldots, y_{2r-3})$ for $y_j \in \mathbb{Z}^d$. Given (T, φ), fix $r - 1$ vertices $\bar{\imath}$ in T. We say that $(T, \varphi, \bar{\imath})$ is *compatible* with $(\sigma; \vec{y}, \vec{m})$ if the following hold:

1. β_B is isomorphic to a subshape λ of σ (in which case the paths of the skeleton B have an induced labelling).
2. Let $l_j > 0$ denote the length of the skeleton path labelled j, and let $l_j = 0$ for any edge in σ that is not in λ. Then $l_j = m_j$ for each $j = 1, \ldots, 2r - 3$.
3. The image under φ of the skeleton path (oriented away from the root) labelled j undergoes the displacement y_j for each j labelling an edge in λ, and $y_j = 0$ for any edge j in σ that is not in λ.

For example, given $(T, \bar{\imath})$ of Fig. 15.3, and any embedding φ of T, $(T, \varphi, \bar{\imath})$ is compatible with $(\sigma_3; (0, \varphi(i_3), \varphi(i_2), \varphi(i_1) - \varphi(i_2), 0), (0, 2, 1, 2, 0))$.

The r-point functions: Let $r \ge 2$, $\sigma \in \Sigma_r$, $\vec{y} = (y_1, \ldots, y_{2r-3})$ with each $y_i \in \mathbb{Z}^d$, and let $\vec{m} = (m_1, \ldots, m_{2r-3})$ with each m_i a non-negative integer. We define

$$b_n^{(r)}(\sigma; \vec{y}, \vec{m}) \tag{15.23}$$
$$= \sum_{(T, \varphi): |T| = n} \mathbb{P}(T, \varphi) \sum_{i_1, \ldots, i_{r-1} \in T} I[(T, \varphi, \bar{\imath}) \text{ is compatible with } (\sigma; \vec{y}, \vec{m})].$$

Then we define the r-point function by

$$b_{z, \zeta}^{(r)}(\sigma; \vec{y}) = \sum_{n=0}^{\infty} \sum_{m_1, \ldots, m_{2r-3}=0}^{\infty} b_n^{(r)}(\sigma; \vec{y}, \vec{m}) z^n \prod_{j=1}^{2r-3} \zeta_j^{m_j}. \tag{15.24}$$

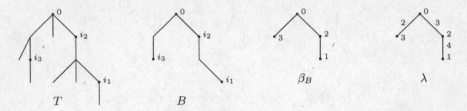

Fig. 15.3. A tree T containing i_1, i_2, i_3, its skeleton B, the reduced skeleton β_B, and the subshape λ of σ_3 (see Fig. 15.1) to which β_B is isomorphic.

Exercise 15.3. Show that (15.24) agrees with the definition (15.16) for $r = 2$.

The next theorem gives the Fourier transform of the r-point functions, where

$$\hat{f}(\vec{k}) = \sum_{y_1,\ldots,y_{2r-3}\in\mathbb{Z}^d} f(\vec{y}) e^{i\vec{k}\cdot\vec{y}} \tag{15.25}$$

with $\vec{k}\cdot\vec{y} = \sum_{j=1}^{2r-3} k_j \cdot y_j$.

Theorem 15.4. *For* $d \geq 1$, $r \geq 2$, $\sigma \in \Sigma_r$, $k_j \in [-\pi,\pi]^d$, $|z| < 1$, $|\zeta_j| \leq 1$,

$$\hat{b}_{z,\zeta}^{(r)}(\sigma;\vec{k}) = \left(b_z^{(1)}\right)^{-2(r-2)} \prod_{j=1}^{2r-3} \hat{b}_{z,\zeta_j}^{(2)}(k_j). \tag{15.26}$$

Before proving the theorem, we remark that the factor $(b_z^{(1)})^{-2(r-2)}$ has a combinatorial interpretation. Namely, it "corrects" for an overcounting of the branch at each of the $r-2$ shape vertices of degree 3, as this branch is counted in $\prod_{j=1}^{2r-3} \hat{b}_{z,\zeta_j}^{(2)}(k_j)$ once by each of the three two-point functions incident at that vertex. This factor is equal to 1 at the critical point $z = 1$, and does not play a role in the leading critical behaviour.

Proof of Theorem 15.4. The statement of the theorem is a tautology for $r = 2$, so we consider $r \geq 3$. Let

$$\hat{q}_{z,\zeta}^{(2)}(k) = \zeta \hat{D}(k) \hat{b}_{z,\zeta}^{(2)}(k). \tag{15.27}$$

By (15.18), $\hat{b}_{z,\zeta}^{(2)}(k) = b_z^{(1)}[1 + \hat{q}_{z,\zeta}^{(2)}(k)]$, so it suffices to show that

$$\hat{b}_{z,\zeta}^{(r)}(\sigma;\vec{k}) = b_z^{(1)} \prod_{j=1}^{2r-3} \left(1 + \hat{q}_{z,\zeta_j}^{(2)}(k_j)\right). \tag{15.28}$$

Expanding the product, the desired identity (15.28) is equivalent to

$$\hat{b}_{z,\zeta}^{(r)}(\sigma;\vec{k}) = b_z^{(1)} \sum_{\lambda \leq \sigma} \prod_{j\in e(\lambda)} \hat{q}_{z,\zeta_j}^{(2)}(k_j). \tag{15.29}$$

Given a subshape $\lambda \leq \sigma$, we let $b(\lambda)$ denote the result of restricting the summation in (15.24) to $m_j = 0$ (and thus $y_j = 0$) for $j \notin e(\lambda)$ and $m_j > 0$ for $j \in e(\lambda)$. Its Fourier transform will be denoted $\hat{b}(\lambda)$. We leave implicit the dependence on the variables of $\vec{k}$ and $\vec{\zeta}$, as these are determined by the edge labels of λ. Then

$$\hat{b}_{z,\zeta}^{(r)}(\sigma; \vec{k}) = \sum_{\lambda \leq \sigma} \hat{b}(\lambda). \qquad (15.30)$$

Thus it suffices to show that

$$\hat{b}(\lambda) = b_z^{(1)} \prod_{j \in e(\lambda)} \hat{q}_{z,\zeta_j}^{(2)}(k_j). \qquad (15.31)$$

This is clear if λ consists of a single vertex, so it suffices to consider the case where λ contains at least one edge.

We use π to denote a subshape for which the root has degree 1, and write $\hat{q}(\pi) = (zp_1)^{-1}\hat{b}(\pi)$. The factor $(zp_1)^{-1}$ serves to cancel the factor $zp_1 = ze^{-1}$ associated to the root in $\hat{b}(\pi)$. Given a subshape λ having at least one edge, let $\pi_1, \ldots, \pi_l$ be the branches emerging from its root. As in (15.18), using now that the lth derivative of P obeys $zP^{(l)}(b_z^{(1)}) = b_z^{(1)}$,

$$\hat{b}(\lambda) = \sum_{j=l}^{\infty} zp_j j(j-1)\cdots(j-l+1)\left(b_z^{(1)}\right)^{j-l} \prod_{a=1}^{l} \hat{q}(\pi_a)$$

$$= b_z^{(1)} \prod_{a=1}^{l} \hat{q}(\pi_a). \qquad (15.32)$$

Let $\bar{\pi}$ denote the subshape obtained from π by contracting the edge incident to the root. We claim that

$$\hat{q}(\pi) = \hat{q}_{z,\zeta}^{(2)}(k)\frac{1}{b_z^{(1)}}\hat{b}(\bar{\pi}), \qquad (15.33)$$

where ζ and k bear the subscript of the label of the edge incident on the root of π. From this, the desired result (15.31) then follows by substituting (15.33) into (15.32) recursively.

To prove (15.33), we condition on whether the length of the tree's skeleton path, corresponding to the edge of π incident on the root, is equal to or greater than 1. This leads, by conditioning as in (15.18), to

$$\hat{q}(\pi) = \zeta\hat{D}(k)\hat{b}(\bar{\pi}) + \zeta\hat{D}(k)b_z^{(1)}\hat{q}(\pi). \qquad (15.34)$$

Solving and using (15.17) and (15.27), we obtain

$$\hat{q}(\pi) = \frac{\zeta\hat{D}(k)}{1 - \zeta\hat{D}(k)b_z^{(1)}}\hat{b}(\bar{\pi}) = \hat{q}_{z,\zeta}^{(2)}(k)\frac{1}{b_z^{(1)}}\hat{b}(\bar{\pi}), \qquad (15.35)$$

which is (15.33). ∎

15.3 Weakly Self-Avoiding Lattice Trees

The weakly self-avoiding walk has played an important role in the development of the theory of self-avoiding walks. There has been no parallel situation for lattice trees, perhaps because it is less obvious how to define weakly self-avoiding lattice trees. In this section we give a natural definition of weakly self-avoiding lattice trees and prove that it corresponds to usual lattice trees in the limit of infinite self-avoidance strength. Throughout the section, we follow the presentation of [30]. Presumably weakly self-avoiding lattice trees are in the same universality class as usual lattice trees, no matter how weak the self-avoidance.

Let $\mathbb{P}(T, \varphi)$ be given by (15.7)–(15.8). Given $\beta \geq 0$, let

$$Z_n^\beta = \sum_{(T,\varphi):|T|=n} \mathbb{P}(T, \varphi) \exp \left[-\tfrac{1}{2}\beta \sum_{i,j \in T: i \neq j} I[\varphi(i) = \varphi(j)] \right], \qquad (15.36)$$

and, for $|T| = n$, define

$$\mathbb{Q}_n^\beta(T, \varphi) = \frac{1}{Z_n^\beta} \mathbb{P}(T, \varphi) \exp \left[-\tfrac{1}{2}\beta \sum_{i,j \in T: i \neq j} I[\varphi(i) = \varphi(j)] \right]. \qquad (15.37)$$

The measure $\mathbb{Q}_n^\beta$ on the set of n-vertex branching random walk configurations rewards self-avoidance by giving a penalty $e^{-\beta}$ to each self-intersection. For $\beta = 0$, $\mathbb{Q}_n^0$ is just branching random walk conditional on $|T| = n$. The next theorem shows that the weakly self-avoiding lattice trees interpolate between branching random walk and lattice trees, in the sense that $\mathbb{Q}_n^\infty$ corresponds in an appropriate sense to the uniform measure on the set of n-vertex lattice trees containing the origin.

In the statement of the theorem $t_n^{(1)}$ denotes the number of n-vertex lattice trees containing the origin, as in Sect. 7.1. Given an injective φ and a lattice tree L, we abuse notation by writing $\varphi(T) = L$ if $\varphi(T)$ consists of the vertices in L and the edges in T are mapped to the bonds in L.

Theorem 15.5. *For $d \geq 1$ and $n \geq 1$,*

$$\lim_{\beta \to \infty} \mathbb{Q}_n^\beta(T, \varphi) = 0 \qquad (15.38)$$

if φ is not injective. On the other hand, given an n-vertex lattice tree L containing the origin,

$$\lim_{\beta \to \infty} \sum_{(T,\varphi):\varphi(T)=L} \mathbb{Q}_n^\beta(T, \varphi) = \frac{1}{t_n^{(1)}}. \qquad (15.39)$$

Proof. The first statement of the theorem, for non-injective φ, follows immediately from the definition of $\mathbb{Q}_n^\beta$.

For the second statement of the theorem, let $\mathcal{T}_n$ denote the set of n-vertex lattice trees containing the origin. This has cardinality $t_n^{(1)}$. We will prove that

$$\sum_{(T,\varphi):\varphi(T)=L} \mathbb{P}(T,\varphi) = (2d)^{-(n-1)}\mathrm{e}^{-n} \qquad (15.40)$$

for every $L \in \mathcal{T}_n$. The important point for the proof is that the right hand side is the same for all $L \in \mathcal{T}_n$, and its particular value plays no role. In fact, given (15.40), we then have

$$Z_n^\infty = \sum_{L \in \mathcal{T}_n} \sum_{(T,\varphi):\varphi(T)=L} \mathbb{P}(T,\varphi) = t_n^{(1)}(2d)^{-(n-1)}\mathrm{e}^{-n}, \qquad (15.41)$$

which gives the desired result that

$$\sum_{(T,\varphi):\varphi(T)=L} \mathbb{Q}_n^\infty(T,\varphi) = \frac{1}{Z_n^\infty} \sum_{(T,\varphi):\varphi(T)=L} \mathbb{P}(T,\varphi) = \frac{1}{t_n^{(1)}}. \qquad (15.42)$$

To prove (15.40), we first note that by (15.7) and (15.8),

$$\sum_{(T,\varphi):\varphi(T)=L} \mathbb{P}(T,\varphi) = (2d)^{-(n-1)}\mathrm{e}^{-n} \sum_{(T,\varphi):\varphi(T)=L} \prod_{i \in T} \frac{1}{\xi_i!}, \qquad (15.43)$$

where ξ_i is the number of offspring of vertex i. It suffices to show that

$$\sum_{(T,\varphi):\varphi(T)=L} \prod_{i \in T} \frac{1}{\xi_i!} = 1. \qquad (15.44)$$

Let b_0 be the degree of 0 in L, and given nonzero $x \in L$, let b_x be the degree of x in L minus 1 (the *forward* degree of x). Then the set $\{b_x : x \in L\}$ (with repetitions) must be equal to the set of ξ_i (with repetitions) for any T that can be mapped to L. Defining $\nu(L)$ to be the cardinality of $\{(T,\varphi) : \varphi(T) = L\}$, (15.44) is therefore equivalent to

$$\nu(L) = \prod_{x \in L} b_x!. \qquad (15.45)$$

We prove (15.45) by induction on the number N of generations of L. By this, we mean the length of the longest self-avoiding path in L, starting from the origin. The identity (15.45) clearly holds if $N = 0$. Our induction hypothesis is that (15.45) holds if there are $N - 1$ or fewer generations. Suppose L has N generations, and let $L_1, \ldots, L_{b_0}$ denote the lattice trees resulting from deleting from L the origin and all bonds incident on the origin. We regard each L_a as rooted at the neighbour of the origin in the corresponding deleted bond. It suffices to show that $\nu(L) = b_0! \prod_{a=1}^{b_0} \nu(L_a)$, since each L_a has fewer than N generations.

To prove this, we note that each pair (T, φ) with $\varphi(T) = L$ induces a set of (T_a, φ_a) such that $\varphi_a(T_a) = L_a$. This correspondence is $b_0!$ to 1, since (T, φ) is determined by the set of (T_a, φ_a), up to permutation of the branches of T at its root. See Exercise 15.6. This proves $\nu(L) = b_0! \prod_{a=1}^{b_0} \nu(L_a)$.   ∎

Exercise 15.6. Let L be the 2-dimensional lattice tree consisting of bonds $\{0, e_2\}$, $\{0, -e_2\}$, $\{0, e_1\}$, $\{e_1, 2e_1\}$, $\{e_1, e_1 + e_2\}$, where $e_1 = (1, 0)$ and $e_2 = (0, 1)$. Let T_1 be the single vertex 0_1 and T_2 be the single vertex 0_2, with $\varphi_1(0_1) = e_2$ and $\varphi_2(0_2) = -e_2$. Let T_3 be the tree consisting of the root 0 and its two offspring 00 and 01, with $\varphi_3(0) = e_1$, $\varphi_3(00) = 2e_1$, $\varphi_3(01) = e_1 + e_2$. Write down the six distinct (T, φ) that correspond to the collection $(T_a; \varphi_a)$ $(a = 1, 2, 3)$ as in the last paragraph of the proof of Theorem 15.5.

16

Integrated Super-Brownian Excursion

In this chapter, we discuss integrated super-Brownian excursion (ISE), and describe how it arises as a scaling limit of branching random walk in all dimensions, and of lattice trees and critical percolation clusters above their upper critical dimensions.

To formulate a scaling limit of branching random walk, it is necessary to decide how much information the limit should keep track of. One possibility, discussed in this chapter, is to condition the branching random walk to have a total of n particles (counting multiple occupations of a vertex and including all generations), to assign mass n^{-1} to each vertex, and to take the scaling limit of this random mass distribution. The mass distribution due to a configuration of branching random walk is a random discrete probability measure on $\mathbb{R}^d$, and we are led naturally to study random probability measures on $\mathbb{R}^d$, which is to say that we study probability measures on probability measures. The scaling limit in this framework is ISE. Much of the theory of ISE was developed in [14, 15, 151]. Our analysis will be based on explicit formulas for the moment measures of ISE, which can be found in [15, 152]. The material in this chapter is based on [30, 61, 105, 188].

The above description of a branching random walk configuration as a mass distribution ignores issues of time evolution and genealogy. As we will see, information about both of these can be recovered, but the focus in this chapter is primarily on mass distribution alone. We will return to the other issues in Chap. 17, where the canonical measure of super-Brownian motion will be discussed.

16.1 Moment Measures of Branching Random Walk

Let $M_1(\mathbb{R}^d)$ denote the set of Borel probability measures on $\mathbb{R}^d$, equipped with the topology of weak convergence. Borel sets in $M_1(\mathbb{R}^d)$ can be defined using this topology, and we consider Borel measures on $M_1(\mathbb{R}^d)$.

Branching random walk configurations (T, φ) with $|T| = n$ induce a measure μ_n on $M_1(\mathbb{R}^d)$, as follows. Let δ_y be the point mass at $y \in \mathbb{R}^d$, i.e., $\delta_y(A) = 1$ if the Borel set A contains y, and otherwise $\delta_y(A) = 0$. Given a tree T with $|T| = n$, and an embedding φ of T, let

$$\mu(T, \varphi) = \frac{1}{n} \sum_{i \in T} \delta_{n^{-1/4}\varphi(i)}. \tag{16.1}$$

Thus $\mu(T, \varphi)$ is the probability measure on $\mathbb{R}^d$ which assigns mass $\frac{1}{n}$ to $n^{-1/4}\varphi(i)$ for each $i \in T$. We then define the measure μ_n on $M_1(\mathbb{R}^d)$ to be the measure that assigns mass $\mathbb{P}((T, \varphi) | |T| = n)$ to each $\mu(T, \varphi)$ with $|T| = n$. In other words, we obtain a random probability measure on $\mathbb{R}^d$ by assigning equal mass to each of the n embedded vertices of a rescaled version of an embedded tree.

We may then ask if the measures μ_n converge weakly to some measure μ_{ISE} on $M_1(\mathbb{R}^d)$. Here, weak convergence (see [24]) is the assertion that for any real-valued bounded continuous function F on $M_1(\mathbb{R}^d)$,

$$\lim_{n \to \infty} \int_{M_1(\mathbb{R}^d)} F(\nu) \mathrm{d}\mu_n(\nu) = \int_{M_1(\mathbb{R}^d)} F(\nu) \mathrm{d}\mu_{\text{ISE}}(\nu). \tag{16.2}$$

A useful ingredient in answering this question is to study the convergence of the moment measures of μ_n. The lth moment measure $M^{(l)}$ of a probability measure μ on $M_1(\mathbb{R}^d)$ is the deterministic probability measure on $(\mathbb{R}^d)^l$ defined, for $l \geq 1$, by

$$dM^{(l)}(x_1, \ldots, x_l) = \int_{M_1(\mathbb{R}^d)} \mathrm{d}\mu(\nu) \mathrm{d}\nu(x_1) \cdots \mathrm{d}\nu(x_l). \tag{16.3}$$

To prove weak convergence of the moment measures, it is sufficient to prove pointwise convergence of their characteristic functions.

To write down the moment measures for μ_n, for $l \geq 1$ we define

$$s_n^{(l+1)}(\bar{x}) = \sum_{(T,\varphi): |T|=n} \mathbb{P}(T, \varphi) \sum_{i_1, \ldots, i_l \in T} \prod_{j=1}^{l} I[\varphi(i_j) = x_j], \tag{16.4}$$

where $\bar{x} = (x_1, \ldots, x_l)$. Note that summation of (16.4) over $\bar{x} \in \mathbb{R}^{dl}$ gives $n^l \mathbb{P}(|T| = n)$. Interpreting (16.3) in this context, the lth moment measure $M_n^{(l)}$ of μ_n is the probability measure on $\mathbb{R}^{dl}$ which places mass equal to $[n^l \mathbb{P}(|T| = n)]^{-1} s_n^{(l+1)}(\bar{x})$ at $n^{-1/4}\bar{x}$, for $\bar{x} \in \mathbb{Z}^{dl}$. The characteristic function $\hat{M}_n^{(l)}(\bar{k})$ of $M_n^{(l)}$ is therefore given by

$$\hat{M}_n^{(l)}(\bar{k}) = \frac{1}{n^l \mathbb{P}(|T| = n)} \hat{s}_n^{(l+1)}(\bar{k} n^{-1/4}), \tag{16.5}$$

where

$$\hat{f}(\bar{k}) = \sum_{\bar{x}} f(\bar{x}) e^{i\bar{k}\cdot\bar{x}} \qquad (16.6)$$

with $\bar{k} \cdot \bar{x} = \sum_{j=1}^{l} k_j \cdot x_j$. We wish to compute the limit of $\hat{M}_n^{(l)}(\bar{k})$ as $n \to \infty$. The asymptotic behaviour of the denominator of (16.5) is given by (15.15) as $(2\pi)^{-1/2} n^{l-3/2}$, so we are left with the numerator.

The limiting behaviour of the numerator will be discussed in Sect. 16.3. It seems clear that there should be a close relationship between the coefficients $\hat{b}_n^{(r)}(\sigma; \bar{k}, \vec{m})$ of (15.23)–(15.24) and $\hat{s}_n^{(r)}(\bar{k})$. In fact, $b_n^{(r)}(\sigma; \vec{y}, \vec{m})$ contains more information than $s_n^{(r)}(\bar{x})$, as it specifies the skeleton displacements in addition to $\bar{x}$. So we first consider the asymptotic behaviour of $\hat{b}_n^{(r)}(\sigma; \bar{k}, \vec{m})$. Since we know the generating functions $\hat{b}_{z,\zeta}^{(r)}(\sigma; \bar{k})$ from Sect. 15.2, this will be studied within the framework of the question: If we know the generating function of a sequence, what can we say about the asymptotic behaviour of the sequence itself? The rule of thumb is that the latter is determined by the singularity of the generating function that is closest to the origin. Moreover, the nature of the singularity can be described in terms of critical exponents.

16.2 Critical Exponents and Generating Functions

In this section, we determine the asymptotic behaviour of $\hat{b}_n^{(r)}(\sigma; \bar{k}, \vec{m})$. To do so, we first study the singularity structure of the generating function $\hat{b}_{z,\zeta}^{(r)}(\sigma; \bar{k})$, beginning with $r = 2$. This holds the key to the study of $r \geq 3$, by Theorem 15.4.

The generating function $\hat{b}_{z,\zeta}^{(2)}(k)$ is given by (15.17), in terms of the one-point function $b_z^{(1)}$ of Theorem 15.1. Properties of the one-point function can be derived from known properties of the Lambert function, or derived directly from (15.12). In particular, $b_z^{(1)}$ has a square root singularity at $z = 1$, and

$$b_z^{(1)} = 1 - \sqrt{2}(1-z)^{1/2} + O(|1-z|), \qquad (16.7)$$

with the absolute value of the error term bounded by a constant multiple of $|1-z|$ uniformly in the cut plane $\mathbb{C} \setminus [1, \infty)$. The square root is the branch with $(1-z)^{1/2} > 0$ for $z < 1$.

Exercise 16.1. Prove (16.7).

Using also the fact that $\hat{D}(k) = 1 - (2d)^{-1}k^2 + O(k^4)$ by (1.12), it follows from (15.17) that

$$\hat{b}_{z,\zeta}^{(2)}(k) = \frac{1 + E_1}{(2d)^{-1}|k|^2 + 2^{1/2}\sqrt{1-z} + (1-\zeta) + E_2}, \qquad (16.8)$$

where $E_1 = -\sqrt{2}(1-z)^{1/2} + O(|1-z|)$ is small compared to 1 for z near 1, and where E_2 is a sum of terms of order $|k|^4$, $|k|^2(1-\zeta)$, $|1-z|$, and similar terms which are higher order than $|k|^2$, $\sqrt{1-z}$ and $(1-\zeta)$.

The formula (16.8) has significance for the values of the critical exponents of branching random walk. Namely, setting $k = 0$ and $\zeta = 1$ in (16.8), we see that $\hat{b}_{z,1}^{(2)}(0) \sim 2^{-1/2}(1-z)^{-1/2}$, and hence the critical exponent γ for the susceptibility of branching random walk is $\gamma = \frac{1}{2}$. Setting $z = 1$ and $\zeta = 1$ in (16.8), we see that $\hat{b}_{1,1}^{(2)}(k) \sim 2d|k|^{-2}$, and hence the critical exponent η for the critical two-point function of branching random walk is $\eta = 0$. Finally, setting $k = 0$ and $z = 1$ gives $\hat{b}_{1,\zeta}^{(2)}(0) \sim (1-\zeta)^{-1}$ (this is, in fact, an equality by (15.17)), so the "backbone exponent" is equal to 1. In addition, information is provided by (16.8) concerning the additive form in the denominator for the joint behaviour as $k \to 0$, $z \to 1$, and $\zeta \to 1$.

The error terms E_1 and E_2 in (16.8) need to be taken into account in an analysis of the moment measures of branching random walk, but this is merely a nuisance and they do not add anything meaningful. In the following, we will simplify the problem by temporarily setting these error terms equal to zero, obtaining

$$\hat{b}_{z,\zeta}^{(2)}(k) \approx \frac{1}{(2d)^{-1}|k|^2 + 2^{1/2}\sqrt{1-z} + (1-\zeta)}. \tag{16.9}$$

Thus, for our simplified version of the two-point function, for $k \in \mathbb{R}^d$ and $\zeta, z \in \mathbb{C}$ with $|\zeta|, |z| < 1$, we define $H_{z,\zeta}^{(2)}(k)$ by

$$H_{z,\zeta}^{(2)}(k) = \frac{1}{(2d)^{-1}|k|^2 + 2^{1/2}\sqrt{1-z} + (1-\zeta)}, \tag{16.10}$$

where the square root has branch cut $[1,\infty)$ and is positive for real $z < 1$. In addition, for $r \geq 2$, $k_j \in \mathbb{R}^d$ and $\zeta_j \in \mathbb{C}$, with $|\zeta_j| < 1$, we write $\vec{k} = (k_1, \ldots, k_{2r-3})$ and $\vec{\zeta} = (\zeta_1, \ldots, \zeta_{2r-3})$, and define

$$H_{z,\vec{\zeta}}^{(r)}(\vec{k}) = \prod_{j=1}^{2r-3} H_{z,\zeta_j}^{(2)}(k_j), \tag{16.11}$$

analogous to (15.26). We write the coefficients in the Taylor expansion of (16.11) as

$$H_{z,\vec{\zeta}}^{(r)}(\vec{k}) = \sum_{m_1,\ldots,m_{2r-3}=0}^{\infty} \sum_{n=0}^{\infty} h_{n,\vec{m}}^{(r)}(\vec{k}) z^n \prod_{j=1}^{2r-3} \zeta_j^{m_j}. \tag{16.12}$$

We write $g_{\vec{m}}^{(r)}(\vec{k}) = \sum_{n=0}^{\infty} h_{n,\vec{m}}^{(r)}(\vec{k})$ for the coefficient of $\prod_{j=1}^{2r-3} \zeta_j^{m_j}$ in $H_{1,\vec{\zeta}}^{(r)}(\vec{k})$. Writing $\vec{1} = (1,\ldots,1)$, we denote the coefficient of z^n in $H_{z,\vec{1}}^{(r)}(\vec{k})$ by

$$h_n^{(r)}(\vec{k}) = \sum_{m_1,\ldots,m_{2r-3}=0}^{\infty} h_{n,\vec{m}}^{(r)}(\vec{k}). \tag{16.13}$$

To state the scaling limits of $h_n^{(r)}(\vec{k})$ and $h_{n,\vec{m}}^{(r)}(\vec{k})$, for $k_i \in \mathbb{R}^d$ and $t_i \geq 0$, we define

$$\hat{a}^{(r)}(\vec{k}, \vec{t}) = \left(\sum_{i=1}^{2r-3} t_i \right) e^{-\left(\sum_{i=1}^{2r-3} t_i\right)^2/2} e^{-\sum_{i=1}^{2r-3} |k_i|^2 t_i/2d} \tag{16.14}$$

and

$$\hat{A}^{(r)}(\vec{k}) = \int_0^\infty dt_1 \cdots \int_0^\infty dt_{2r-3} \hat{a}^{(r)}(\vec{k}, \vec{t}). \tag{16.15}$$

Explicitly, for $r = 2$,

$$\hat{A}^{(2)}(k) = \int_0^\infty t e^{-t^2/2} e^{-|k|^2 t/2d} \, dt. \tag{16.16}$$

The integral (16.16) can be written in terms of the parabolic cylinder function D_{-2} as $\hat{A}^{(2)}(\sqrt{d}k) = e^{k^4/16} D_{-2}(|k|^2/2)$ [83, 3.462.1]. The integrals (16.15) obey

$$\hat{A}^{(r)}(\vec{0}) = \frac{1}{(2r-5)!!}, \tag{16.17}$$

where we recall the notation of (15.22), see Exercise 16.4.

We write $\lfloor \vec{t}n \rfloor$ to denote the vector with components $\lfloor t_j n \rfloor$.

Proposition 16.2. *For $r \geq 2$, as $n \to \infty$,*

$$g_{\lfloor \vec{t}n \rfloor}^{(r)}(\vec{k}n^{-1/2}) \to e^{-\sum_{j=1}^{2r-3} |k_j|^2 t_j/2d}, \tag{16.18}$$

$$h_n^{(r)}(\vec{k}n^{-1/4}) \sim \frac{1}{\sqrt{2\pi}} n^{r-5/2} \hat{A}^{(r)}(\vec{k}), \tag{16.19}$$

$$h_{n,\lfloor \vec{t}n^{1/2} \rfloor}^{(r)}(\vec{k}n^{-1/4}) \sim \frac{1}{\sqrt{2\pi}} \frac{1}{n} \hat{a}^{(r)}(\vec{k}, \vec{t}). \tag{16.20}$$

Proof. For (16.18), we use the fact that $H_{1,\zeta}^{(2)}(k)$ is the sum of a geometric series in ζ, namely

$$H_{1,\zeta}^{(2)}(k) = \frac{1}{(2d)^{-1}|k|^2 + (1 - \zeta)} = \sum_{m=0}^\infty \frac{1}{(1 + |k|^2/2d)^{m+1}} \zeta^m. \tag{16.21}$$

Therefore $g_{\vec{m}}^{(r)}(\vec{k}) = \prod_{j=1}^{2r-3} (1 + |k_j|^2/2d)^{-(m_j+1)}$, and (16.18) follows.

For the remainder of the proof, we follow [62], and to simplify the formulas we put $d = 1$ (the general case then easily follows). For (16.19), it follows from the Cauchy integral formula that

$$h_n^{(r)}(\vec{k}n^{-1/4}) = \frac{1}{2\pi i} \oint \frac{dz}{z^{n+1}} \prod_{j=1}^{2r-3} \frac{2}{|k_j|^2 n^{-1/2} + 2^{3/2}(1-z)^{1/2}}, \tag{16.22}$$

where the integral is taken around a small circle centred at the origin. For simplicity, in the proof of (16.19) we suppose that $k_j \neq 0$ for all j. See Exercise 16.3 for the general case. We make the change of variables $w = n(z - 1)$, and then deform the contour to the branch cut $[0, \infty)$ in the w-plane. The resulting contour goes from right to left below the branch cut and from left to right above the cut. We then apply the identity

$$\frac{2}{|k_j|^2 + 2^{3/2}(-w)^{1/2}} = \int_0^\infty dt_j\, e^{-|k_j|^2 t_j/2} e^{-2^{1/2}(-w)^{1/2} t_j}. \qquad (16.23)$$

Taking into account the correct branch of the square root on either side of the branch cut, and applying Fubini's theorem, this gives

$$h_n^{(r)}(\vec{k}n^{-1/4}) = n^{r-5/2} \int_0^\infty dt_1 \cdots \int_0^\infty dt_{2r-3}\, e^{-(\sum_{j=1}^{2r-3} |k_j|^2 t_j)/2}$$

$$\times \frac{1}{\pi} \int_0^\infty \frac{dw}{(1+w/n)^{n+1}} \sin((\textstyle\sum_{j=1}^{2r-3} t_j)\sqrt{2w}). \qquad (16.24)$$

Since $(1 + \frac{w}{n})^{n+1} \geq 1 + \frac{(n+1)n}{2}(\frac{w}{n})^2 \geq 1 + \frac{w^2}{2}$ for all $n \geq 1$, the dominated convergence theorem can be applied to conclude that as $n \to \infty$ the w-integral converges to

$$\int_0^\infty dw\, e^{-w} \sin((\textstyle\sum_{j=1}^{2r-3} t_j)\sqrt{2w}) = \left(\frac{\pi}{2}\right)^{1/2} \left(\sum_{j=1}^{2r-3} t_j\right) e^{-(\sum_{j=1}^{2r-3} t_j)^2/2}.$$

$$\qquad (16.25)$$

Therefore, as required,

$$h_n^{(r)}(\vec{k}n^{-1/4}) \sim \frac{1}{\sqrt{2\pi}} n^{r-5/2} \hat{A}^{(r)}(\vec{k}). \qquad (16.26)$$

Next, we prove (16.20) (again taking $d = 1$ for simplicity). Since each factor in the product in (16.11) is the sum of a geometric series in ζ_j,

$$h_{n,\lfloor \vec{t}n^{1/2}\rfloor}^{(r)}(\vec{k}n^{-1/4})$$

$$= \frac{1}{2\pi i} \oint \frac{dz}{z^{n+1}} \prod_{j=1}^{2r-3} \left(1 + \frac{|k_j|^2}{2\sqrt{n}} + \sqrt{2}(1-z)^{1/2}\right)^{-(\lfloor t_j n^{1/2}\rfloor +1)}, \qquad (16.27)$$

where the integration is performed around a small circle centred at the origin. Making the change of variables $w = n(z - 1)$ and deforming the contour of integration to the branch cut $[0, \infty)$ in the w-plane, the above is equal to

$$\frac{1}{\pi n} \mathrm{Im} \int_0^\infty \frac{dw}{(1+w/n)^{n+1}} \prod_{j=1}^{2r-3} \left(1 + \frac{|k_j|^2}{2\sqrt{n}} - \frac{i\sqrt{2w}}{\sqrt{n}}\right)^{-(\lfloor t_j n^{1/2}\rfloor +1)}. \qquad (16.28)$$

The integrand here is dominated by $(1 + \frac{w^2}{2})^{-1} \cdot 1$, so by the dominated convergence theorem, this is asymptotic to

$$\frac{1}{\pi n} e^{-(\sum_{j=1}^{2r-3} |k_j|^2 t_j)/2} \int_0^\infty dw\, e^{-w} \sin((\textstyle\sum_{j=1}^{2r-3} t_j)\sqrt{2w}) = \frac{1}{\sqrt{2\pi n}} \hat{a}^{(r)}(\vec{k}, \vec{t}),$$

$$(16.29)$$

using (16.25) in the last step. ∎

Exercise 16.3. Prove (16.19) in the degenerate case where one or more k_j is zero.

Exercise 16.4. Compute the asymptotic behaviour of $h_n^{(r)}(\vec{0})$ directly from its definition, and conclude (16.17) from (16.19).

By taking a little additional care with the error terms E_1 and E_2 in (16.8), a similar analysis shows that for $r \geq 2$, as $n \to \infty$,

$$\sum_{\vec{m}} \hat{b}_n^{(r)}(\sigma; \vec{k}n^{-1/4}, \vec{m}) \sim \frac{1}{\sqrt{2\pi}} n^{r-5/2} \hat{A}^{(r)}(\vec{k}), \qquad (16.30)$$

$$\hat{b}_n^{(r)}(\sigma; \vec{k}n^{-1/4}, \lfloor \vec{t}n^{1/2} \rfloor) \sim \frac{1}{\sqrt{2\pi}} \frac{1}{n} \hat{a}^{(r)}(\vec{k}, \vec{t}). \qquad (16.31)$$

The implications of this for the scaling limit of branching random walk will be discussed in the next section. For now, we note that in (16.31) the scaling of time (distance along the skeleton) is $n^{1/2}$. Since space is being scaled as $n^{1/4}$, this is Brownian scaling. Thus (16.31) provides an interpretation of the variables t_j in $\hat{a}^{(r)}(\vec{k}, \vec{t})$ as rescaled Brownian time variables along skeleton paths. The sum over $\vec{m}$ in (16.30) can be interpreted as a Riemann sum approximation to the integral defining $\hat{A}^{(r)}(k)$.

16.3 Construction of ISE

In this section, we present a proof that rescaled critical Poisson branching random walk converges to ISE, constructing ISE in the process. The proof is based on [30]. We will also derive explicit formulas for the characteristic functions of the ISE moment measures. Our proof applies directly only to the Poisson case, due to its reliance on the particular form of the generating functions derived in Theorems 15.1, 15.2 and 15.4. However, the theorem actually holds more generally, whenever the random variable ξ for the offspring distribution has a finite second moment [14,15].

As an important first step, we prove convergence of the moment measures of the probability measures μ_n on $M_1(\mathbb{R}^d)$ defined in Sect. 16.1.

Theorem 16.5. For $d \geq 1$ and $l \geq 1$, the moment measures $M_n^{(l)}$ of μ_n converge weakly to limiting measures $M^{(l)}$.

Proof. It suffices to prove that for $\bar{k} \in \mathbb{R}^{dl}$ the limits

$$\lim_{n \to \infty} \hat{M}_n^{(l)}(\bar{k}) = \hat{M}^{(l)}(\bar{k}) \qquad (16.32)$$

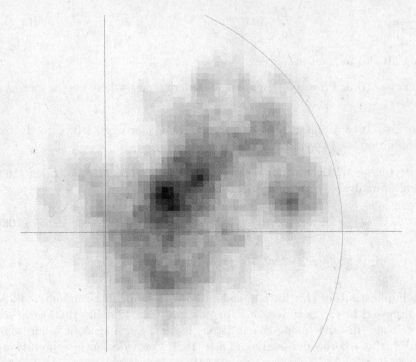

Fig. 16.1. A random mass distribution drawn from μ_n, with $d = 2$ and $n \approx 96,000$. Darker shading represents higher mass, and the arc has radius $n^{1/4}$.

exist for all $l \geq 1$, with $\hat{M}^{(l)}$ continuous at $\bar{0}$. By (16.5) and (15.15), the characteristic function of $M_n^{(l)}$ is

$$\hat{M}_n^{(l)}(\bar{k}) = \frac{1}{n^l \mathbb{P}(|T| = n)} \hat{s}_n^{(l+1)}(\bar{k} n^{-1/4})$$
$$\sim \sqrt{2\pi} n^{-(l-3/2)} \hat{s}_n^{(l+1)}(\bar{k} n^{-1/4}), \qquad (16.33)$$

where $s_n^{(l+1)}$ is given by (16.4).

Define $b_n^{(r)}(\sigma; \vec{y})$, for $r \geq 2$, by

$$b_n^{(r)}(\sigma; \vec{y}) = \sum_{\vec{m}} b_n^{(r)}(\sigma; \vec{y}, \vec{m}), \qquad (16.34)$$

where the summand is defined in (15.23). For $r = 2, 3$, there is only one shape and, suppressing σ in the notation, we have

$$s_n^{(2)}(x) = b_n^{(2)}(x), \quad s_n^{(3)}(x_1, x_2) = \sum_{y \in \mathbb{Z}^d} b_n^{(3)}(y, x_1 - y, x_2 - y). \qquad (16.35)$$

It then follows from (16.30) that

$$\lim_{n\to\infty} \hat{M}_n^{(1)}(k) = \hat{A}^{(2)}(k) = \hat{M}^{(1)}(k) \tag{16.36}$$

and

$$\lim_{n\to\infty} \hat{M}_n^{(2)}(k_1, k_2) = \hat{A}^{(3)}(k_1 + k_2, k_1, k_2) = \hat{M}^{(2)}(k_1, k_2), \tag{16.37}$$

where $\hat{M}^{(1)}(k)$ and $\hat{M}^{(2)}(k_1, k_2)$ are defined by the second equalities. These are continuous at zero by definition.

The situation is more subtle for $l \geq 3$. To begin, we define a way of associating a $2l - 1$ component vector $\vec{k}$ to an l component vector $\bar{k}$, generalizing the expression $\vec{k} = (k_1 + k_2, k_1, k_2)$ in (16.37) for $l = 2$. This is done as follows. Given a shape $\sigma \in \Sigma_{l+1}$, for each vertex j of degree 1 in σ, other than vertex 0, we let ω_j be the set of edges in σ on the path from 0 to j $(j = 1, \ldots, l)$. Given $\bar{k} = (k_1 \ldots, k_l) \in \mathbb{R}^{dl}$, we define $\vec{k}(\sigma) \in \mathbb{R}^{(2l-1)d}$ by setting its ith component $\vec{k}_i(\sigma) \in \mathbb{R}^d$, for $i = 1, \ldots, 2l - 1$, to be

$$\vec{k}_i(\sigma) = \sum_{j=1}^{l} k_j I[i \in \omega_j], \tag{16.38}$$

where, on the right hand side, k_j denotes the jth component of $\bar{k}$ and I is an indicator function. For example, for $l = 3$ and the shape σ_1 of Fig. 15.1, $\vec{k}(\sigma_1) = (k_1 + k_2 + k_3, k_1, k_2 + k_3, k_2, k_3)$. We will prove (16.32) with

$$\hat{M}^{(l)}(\bar{k}) = \sum_{\sigma \in \Sigma_{l+1}} \hat{A}^{(l+1)}(\vec{k}(\sigma)), \tag{16.39}$$

which is continuous at $\bar{0}$ by definition. The limit (16.39) has been established already for $l = 1, 2$.

For $l \geq 3$, we first note that (15.15) and (16.30) imply that

$$\lim_{n\to\infty} \frac{\sum_{\sigma \in \Sigma_{l+1}} \hat{b}_n^{(l+1)}(\sigma; \vec{k}n^{-1/4})}{n^l \mathbb{P}(|T| = n)} = \sum_{\sigma \in \Sigma_{l+1}} \hat{A}^{(l+1)}(\vec{k}). \tag{16.40}$$

Let

$$e_n^{(l+1)}(\bar{k}) = \hat{s}_n^{(l+1)}(\bar{k}) - \sum_{\sigma \in \Sigma_{l+1}} \hat{b}_n^{(l+1)}(\sigma; \vec{k}(\sigma)). \tag{16.41}$$

In view of (16.33) and (16.40), to prove convergence of the lth moments, for $l \geq 3$, it suffices to show that

$$|e_n^{(l+1)}(\bar{k})| \leq O(n^{l-2}). \tag{16.42}$$

The remainder of the proof is devoted to obtaining (16.42).

Recall the definition of $\omega_j = \omega_j(\sigma)$ above (16.38). For $\sigma \in \Sigma_{l+1}$, let

$$Y_{\bar{x}}(\sigma) = \{\vec{y} \in \mathbb{R}^{d(2l-1)} : \sum_{i \in \omega_j(\sigma)} y_i = x_j \; \forall j = 1, \ldots, l\}. \tag{16.43}$$

Recall the definitions above (15.23), and note that if $(T, \varphi, \bar{\imath})$ is compatible with $(\sigma; \vec{y}, \vec{m})$, then the statements (i) $\varphi(i_j) = x_j$ for all $j = 1, \ldots, l$, and (ii) $\vec{y} \in Y_{\bar{x}}(\sigma)$, are equivalent. Consequently,

$$e_n^{(l+1)}(\bar{k}) = \sum_{\bar{x}} e^{i\bar{k} \cdot \bar{x}} \left[s_n^{(l+1)}(\bar{x}) - \sum_{\sigma \in \Sigma_{l+1}} \sum_{\vec{y} \in Y_{\bar{x}}(\sigma)} \sum_{\vec{m}} b_n^{(l+1)}(\sigma; \vec{y}, \vec{m}) \right]$$

$$= \sum_{\bar{x}} e^{i\bar{k} \cdot \bar{x}} \sum_{(T, \varphi) : |T| = n} \mathbb{P}(T, \varphi) \sum_{i_1, \ldots, i_l \in T} \left[\prod_{j=1}^{l} I[\varphi(i_j) = x_j] \right]$$

$$\times \left[1 - \sum_{\sigma \in \Sigma_{l+1}} \sum_{\vec{y} \in Y_{\bar{x}}(\sigma)} \sum_{\vec{m}} I[(T, \varphi, \bar{\imath}) \text{ comp. with } (\sigma; \vec{y}, \vec{m})] \right].$$
$$\tag{16.44}$$

The last line of the above expression need not be zero. For example, let T be the unique tree having just two vertices, let φ be the embedding of T which maps the root of T to 0 and the other vertex of T to a neighbour e_1 of the origin, let i_1 and i_2 be the root of T, and let i_3 be the other vertex of T. In this example, the subtracted sum in (16.44) is equal to 3, due to the terms with $\sigma = \sigma_1$, $y_1 = (0, 0, 0, 0, e_1)$; $\sigma = \sigma_2$, $y_2 = (0, 0, 0, 0, e_1)$; and $\sigma = \sigma_3$, $y_3 = (0, e_1, 0, 0, 0)$, where the shapes σ_i are given in Fig. 15.1.

However, given $(T, \varphi, \bar{\imath})$, and given $\sigma \in \Sigma_{l+1}$, if $(T, \varphi, \bar{\imath})$ is compatible with $(\sigma; \vec{y}, \vec{m})$ then $\vec{y}$ and $\vec{m}$ are uniquely determined by $(T, \varphi, \bar{\imath})$ and σ. Thus for each σ contributing to the last line of (16.44), there is exactly one $\vec{y}$ and one $\vec{m}$ that contribute to the sums over $\vec{y}$ and $\vec{m}$. Also, each $(T, \varphi, \bar{\imath})$, such that the reduced skeleton $\beta_B(T, \bar{\imath})$ of $(T, \bar{\imath})$ is isomorphic to a shape in Σ_{l+1} (rather than to a strict subshape), receives a contribution to (16.44) due to that shape and only to that shape. Therefore, writing Λ_σ for the set of all strict subshapes of $\sigma \in \Sigma_{l+1}$,

$$|e_n^{(l+1)}(\bar{k})| \leq \sum_{\sigma \in \Sigma_{l+1}} \sum_{\lambda \in \Lambda_\sigma} \sum_{\bar{x}} \sum_{(T, \varphi) : |T| = n} \mathbb{P}(T, \varphi)$$

$$\times \sum_{i_1, \ldots, i_l \in T} \left[\prod_{j=1}^{l} I[\varphi(i_j) = x_j] \right] I[\beta_B(T, \bar{\imath}) \simeq \lambda]$$

$$= \sum_{\sigma \in \Sigma_{l+1}} \sum_{\lambda \in \Lambda_\sigma} \sum_{(T, \varphi) : |T| = n} \mathbb{P}(T, \varphi) \sum_{i_1, \ldots, i_l \in T} I[\beta_B(T, \bar{\imath}) \simeq \lambda]. \tag{16.45}$$

Let $f_n^{(l+1)}$ denote the right hand side of (16.45). It suffices to show that $f_n^{(l+1)} \leq O(n^{l-2})$. For this, we introduce the generating function $F^{(l+1)}(z) =$

$\sum_{n=1}^{\infty} f_n^{(l+1)} z^n$. This is a sum of terms of the form $\hat{b}(\lambda)$ (defined below (15.29)), where λ is a strict subshape and all $k_j = 0$, $\zeta_j = 1$. By (15.31), (15.27) and (16.8),

$$|\hat{b}(\lambda)| \leq O(1) \prod_{j \in e(\lambda)} \frac{1}{|1-z|^{1/2}}, \tag{16.46}$$

and hence $|F^{(l+1)}(z)| \leq O(|1-z|^{-(l-1)})$ uniformly in $|z| < 1$, where the power $l - 1 = \frac{1}{2}(2l - 2)$ arises because at least one of the $2l - 1$ skeleton paths is contracted. Then Exercise 7.7 implies the desired bound $f_n^{(l+1)} \leq O(n^{l-2})$. ∎

Using Theorem 16.5, we can now prove convergence of μ_n to a limiting measure μ_{ISE} on $M_1(\mathbb{R}^d)$, thereby constructing ISE. We will use the fact that $M_1(\mathbb{R}^d)$ is a metric space (in fact it is a Polish space) whose metric induces the weak topology; references for this are given in [177, p. 148].

Theorem 16.6. *For $d \geq 1$, there is a measure μ_{ISE} on $M_1(\mathbb{R}^d)$ such that μ_n converges weakly to μ_{ISE} on $M_1(\mathbb{R}^d)$.*

Proof. We will prove that μ_n is relatively compact, i.e., that every subsequence of μ_n contains a weakly convergent subsequence (see [24]). Given a subsequence of μ_n, relative compactness implies that some further subsequence converges to a limit. In principle, this limit could depend on the subsequence. But Theorem 16.5 implies that the limit has moment measures $M^{(l)}$, which determines it uniquely.[1] We call the common limit μ_{ISE}.

To prove relative compactness, we apply Theorem II.4.1 of [177]. This theorem actually applies in the more general context of convergence of measure-valued processes. We have no time-dependence here, so our processes are constant. Condition (ii) of Theorem II.4.1 is vacuous for our purpose.[2] It suffices to verify Condition (i) of Theorem II.4.1, which asserts that given $\epsilon > 0$ there is a compact set K_ϵ such that

$$\sup_n \mu_n(\nu(K_\epsilon^c) > \epsilon) < \epsilon. \tag{16.47}$$

Take $K_\epsilon = B_R = \{x \in \mathbb{R}^d : |x| \leq R\}$, where we will choose $R = R(\epsilon)$ below. Let ψ_R be a smooth function which takes values in $[0, 1]$, which is 1 on B_R^c, and which is 0 on B_{R-1}. Then

$$\mu_n(\nu(K_\epsilon^c) > \epsilon) \leq \epsilon^{-1} \mathbb{E}_{\mu_n}[\nu(K_\epsilon^c)]$$
$$\leq \epsilon^{-1} \mathbb{E}_{\mu_n}[\nu(\psi_R)]$$
$$= \epsilon^{-1} \int_{\mathbb{R}^d} \psi_R(x) \mathrm{d}M_n^{(1)}(x). \tag{16.48}$$

---

[1] It is a consequence of [60, Lemma 2.4.1] that moment measures uniquely determine a measure on $M_1(\dot{\mathbb{R}}^d)$, where $\dot{\mathbb{R}}^d$ is the one-point compactification of $\mathbb{R}^d$. Since the moments have no mass at infinity, they therefore uniquely determine a measure on $M_1(\mathbb{R}^d)$.

[2] We take $e^{ik \cdot x}$ as our separating class and observe that its integral with respect to a probability measure ν is bounded by 1, so $\nu(e^{ik \cdot x})$ is relatively compact.

By Theorem 16.5,

$$\lim_{n\to\infty} \int_{\mathbb{R}^d} \psi_R(x) \mathrm{d}M_n^{(1)}(x) = \int_{\mathbb{R}^d} \psi_R(x) \mathrm{d}M^{(1)}(x)$$
$$\leq \int_{|x|>R-1} \mathrm{d}M^{(1)}(x). \qquad (16.49)$$

Since $\int_{\mathbb{R}^d} \mathrm{d}M^{(1)}(x) = 1$, the right hand side is less than $\epsilon^2/2$ if we choose R large depending on ϵ, and therefore the integral on the left hand side is less than ϵ^2 for this R if we choose n large depending on R and hence on ϵ, say $n \geq n_0(\epsilon)$. We have proved that

$$\sup_{n \geq n_0(\epsilon)} \mu_n(\nu(K_\epsilon^c) > \epsilon) < \epsilon, \qquad (16.50)$$

and we are left with $n < n_0(\epsilon)$.

Note that (16.50) continues to hold if we increase R, since this only decreases the left hand side. However, if we now increase R (if necessary) until it exceeds $n_0(\epsilon)$, then $\nu(K_\epsilon^c) = 0$ almost surely under μ_n for any $n < n_0(\epsilon)$, since under μ_n it is the case that ν is supported on the ball in $\mathbb{R}^d$ of radius $n^{1-1/4} = n^{3/4}$. Thus (16.50) implies (16.47). ∎

16.4 Lattice trees and ISE

It was conjectured by Aldous [15] that ISE describes the scaling limit of lattice trees in dimensions $d > 8$. In this section, we discuss the theorem of [62] that this is the case for sufficiently spread-out lattice trees, following the presentation of [188].

Basic definitions for lattice trees were given in Sect. 7.1. In particular, $t_n^{(1)}$ was defined to be the number of n-bond lattice trees that contain the origin, and $t_n^{(2)}(x)$ was defined to be the number of n-bond lattice trees that contain the origin and x. Now we also define functions $t_n^{(r)}(\sigma; \vec{y}, \vec{s})$ for $r \geq 2$, using the terminology introduced above Theorem 15.4 for branching random walk, which we recast here in a form appropriate for lattice trees.

Given a lattice tree T containing the vertices $0, x_1, \ldots, x_{r-1}$ (not necessarily distinct), we define the *skeleton* B to be the subtree of T spanning $0, x_1, \ldots, x_{r-1}$. Let β_B be the tree (not a lattice tree) obtained by ignoring vertices in B of degree 2, other than $0, x_1, \ldots, x_{r-1}$. Let $\sigma \in \Sigma_r$, $\vec{m} = (m_1, \ldots, m_{2r-3})$ for non-negative integers m_j, and $\vec{y} = (y_1, \ldots, y_{2r-3})$ for $y_j \in \mathbb{Z}^d$. We say that $(T; \bar{x})$ is *compatible* with $(\sigma; \vec{y}, \vec{m})$ if the following hold:

1. β_B is isomorphic to a subshape λ of σ (in which case the paths of the skeleton B have an induced labelling).
2. Let $l_j > 0$ denote the length of the skeleton path labelled j, and let $l_j = 0$ for any edge in σ that is not in λ. Then $l_j = m_j$ for each $j = 1, \ldots, 2r-3$.

3. The skeleton path (oriented away from the origin) labelled j undergoes the displacement y_j for each j labelling an edge in λ, and $y_j = 0$ for any edge j in σ that is not in λ.

Then we define $t_n^{(r)}(\sigma; \vec{y}, \vec{m})$ to be the number of n-bond lattice trees T, containing the origin, for which there are vertices $x_1, \ldots, x_{r-1} \in T$ such that $(T; \bar{x})$ is compatible with $(\sigma; \vec{y}, \vec{m})$. (Note that $\bar{x}$ is in fact uniquely determined by σ and $\vec{y}$.) See Fig. 16.2. We also define

$$t_n^{(r)}(\sigma; \vec{y}) = \sum_{\vec{m}} t_n^{(r)}(\sigma; \vec{y}, \vec{m}), \tag{16.51}$$

where the sum over $\vec{m}$ denotes a sum over the non-negative integers m_j. For $r = 2, 3$ there is a unique shape and we sometimes omit it from the notation. We also define the generating functions

$$G_{z,\vec{\zeta}}^{(r)}(\sigma; \vec{y}) = \sum_{n=0}^{\infty} \sum_{\vec{m}} t_n^{(r)}(\sigma; \vec{y}, \vec{m}) z^n \prod_{j=1}^{2r-3} \zeta_j^{m_j}, \tag{16.52}$$

for $r \geq 2$. For $r \geq 3$, these generating functions keep track of more information than those defined in (7.20). The sum over $\vec{y} \in \mathbb{R}^{d(2r-3)}$ of (16.52) is finite for $|z| < z_c$ and $|\zeta_j| \leq 1$, for all r.

The critical exponents η and γ were defined in (7.8) and (7.6) by

$$\hat{G}_{z_c,1}^{(2)}(k) \sim \text{const} \frac{1}{|k|^{2-\eta}} \quad \text{as } k \to 0, \quad \hat{G}_{z,1}^{(2)}(0) \sim \text{const} \frac{1}{(1 - z/z_c)^{\gamma}} \quad \text{as } z \to z_c. \tag{16.53}$$

Their mean-field values are $\eta = 0$ and $\gamma = \frac{1}{2}$. By analogy with (16.9), we might hope and expect that in dimensions $d > 8$,

$$\hat{G}_{z,\zeta}^{(2)}(k) = \frac{C_0}{D_0^2|k|^2 + 2^{3/2}(1 - z/z_c)^{1/2} + 2T_0(1 - \zeta)} + \text{error}, \tag{16.54}$$

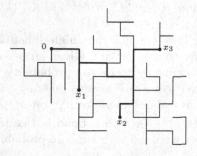

Fig. 16.2. A 2-dimensional lattice tree contributing to $t_{78}^{(4)}(\sigma_1; \vec{y}, \vec{m})$, with σ_1 depicted in Fig. 15.1, $\vec{y} = ((2, -1), (0, -2), (4, -1), (-1, -3), (2, 2))$, $\vec{m} = (3, 2, 5, 4, 4)$.

where C_0, D_0, T_0 are positive constants, and where the error term is of lower order than the main term in some suitable sense, as $k \to 0$, $z \to z_c$, and $\zeta \to 1$. Furthermore, by analogy with Theorem 15.4, we might expect that for $d > 8$ there is approximate independence of the form

$$\hat{G}_{z,\zeta}^{(r)}(\sigma; \vec{k}) = V^{r-2} \prod_{j=1}^{2r-3} \hat{G}_{z,\zeta_j}^{(2)}(k_j) + \text{error}, \tag{16.55}$$

where V is a finite positive constant which encompasses the self-avoidance interactions of lattice trees in a renormalized vertex factor (cf. (7.41), and also Sect. 5.3 and Theorems 6.9 and 6.10).

For the nearest-neighbour model with d sufficiently large, and for spread-out models for $d > 8$ with L sufficiently large, relations of the form (16.54) and (16.55) are proved in [62], for all $r \geq 2$ if $\vec{\zeta} = \vec{1}$ and for $r = 2, 3$ for general $\vec{\zeta}$. The results given below arise as a consequence. For the statement of the results, we define

$$p_n^{(r)}(\sigma; \vec{y}) = \frac{t_n^{(r)}(\sigma; \vec{y})}{\sum_{\sigma \in \Sigma_m} \hat{t}_n^{(r)}(\sigma; \vec{0})}, \tag{16.56}$$

which is a probability measure on $\Sigma_r \times \mathbb{Z}^{d(2r-3)}$. The following theorem from [61, 62], whose proof extends the methods of [95, 99], shows that (16.56) has the corresponding ISE density as its scaling limit in high dimensions. In its statement, the scaling of $\vec{k}$ corresponds to scaling down the lattice spacing by a multiple of $n^{-1/4}$.

Theorem 16.7. *Let $r \geq 2$, $\sigma \in \Sigma_r$, $k_j \in \mathbb{R}^d$ $(j = 1, \ldots, 2r - 3)$, and let D_0 be given by (7.18). For nearest-neighbour lattice trees in sufficiently high dimensions $d \geq d_0$, and for spread-out lattice trees with $d > 8$ and L sufficiently large depending on d, there is a constant[3] c_1 depending on d and L such that as $n \to \infty$*

$$\hat{t}_n^{(r)}(\sigma; \vec{k} D_0^{-1} n^{-1/4}) \sim c_1 n^{r-5/2} z_c^{-n} \hat{A}^{(r)}(\vec{k}). \tag{16.57}$$

In particular,

$$\lim_{n\to\infty} \hat{p}_n^{(r)}(\sigma; \vec{k} D_0^{-1} n^{-1/4}) = \hat{A}^{(r)}(\vec{k}). \tag{16.58}$$

It is a corollary of Theorem 16.7 that high-dimensional lattice trees converge weakly to ISE, as we now explain. Given an n-bond lattice tree T containing the origin, we define ν_T to be the probability measure on $\mathbb{R}^d$ which for each of the $n + 1$ vertices $x \in T$ assigns mass $(n + 1)^{-1}$ to $xD_0^{-1}n^{-1/4}$ in $\mathbb{R}^d$. We then define a probability measure μ_n^{LT} on $M_1(\mathbb{R}^d)$, supported on the ν_T, by $\mu_n^{\text{LT}}(\{\nu_T\}) = (t_n^{(1)})^{-1}$ for each n-bond lattice tree T containing 0. In this way, n-bond lattice trees induce a random probability measure on $\mathbb{R}^d$.

---

[3] The constant c_1 is equal to the constant A of Theorems 7.3 and 7.4, but A is already in use with a different meaning in (16.57).

Corollary 16.8. *For nearest-neighbour lattice trees in sufficiently high dimensions $d \geq d_0$, and for spread-out lattice trees with $d > 8$ and L sufficiently large, μ_n^{LT} converges weakly to μ_{ISE}, as measures on $M_1(\mathbb{R}^d)$.*

The argument leading from Theorem 16.7 to Corollary 16.8 is a straightforward adaptation of the proof of Theorem 16.6. The details are made explicit in [188, Appendix A], where weak convergence of μ_n^{LT} to μ_{ISE} is proved.[4]

For a more refined statement than Theorem 16.7, we define

$$p_n^{(r)}(\sigma; \vec{y}, \vec{m}) = \frac{t_n^{(r)}(\sigma; \vec{y}, \vec{m})}{\sum_{\sigma \in \Sigma_r} \hat{t}_n^{(r)}(\sigma; \vec{0})}, \qquad (16.59)$$

which is a probability measure on $\Sigma_r \times \mathbb{Z}^{d(2r-3)} \times \mathbb{Z}_+^{2r-3}$. The following theorem is due to [61, 62]. In its statement we drop σ, since there is a unique shape for $r = 2, 3$.

Theorem 16.9. *Let $r = 2$ or $r = 3$, $k_j \in \mathbb{R}^d$, and $t_j \in (0, \infty)$ $(j = 1, \ldots, 2r - 3)$. For nearest-neighbour lattice trees in sufficiently high dimensions $d \geq d_0$, and for spread-out lattice trees with $d > 8$ and L sufficiently large depending on d, there is a constant T_0 depending on d and L, such that as $n \to \infty$,*

$$\hat{t}_n^{(r)}(\vec{k} D_0^{-1} n^{-1/4}, \lfloor \vec{t} T_0 n^{1/2} \rfloor) \sim c_1 T_0^{-(2r-3)} n^{-1} z_c^{-n} \hat{a}^{(r)}(\vec{k}, \vec{t}), \qquad (16.60)$$

where c_1 and D_0 are the constants of Theorem 16.7. In particular,

$$\lim_{n \to \infty} (T_0 n^{1/2})^{2r-3} \hat{p}_n^{(r)}(\vec{k} D_0^{-1} n^{-1/4}, \lfloor \vec{t} T_0 n^{1/2} \rfloor) = \hat{a}^{(r)}(\vec{k}, \vec{t}). \qquad (16.61)$$

It would be very surprising if Theorem 16.9 did not extend to all $r \geq 2$, but technical difficulties arise for $r \geq 4$ and the theorem has been proved[5] only for $r = 2$ and $r = 3$. Theorem 16.9 indicates that, at least for $r = 2$ and $r = 3$, skeleton paths with length of order $n^{1/2}$ are typical. This is Brownian scaling, since distance is scaled as $n^{1/4}$, and the variables t_j correspond to rescaled skeleton time variables.

See Sect. 17.5 for further results in this direction. See also [182] for a study of the time evolution of ISE.

16.5 Critical Percolation and ISE

In this section, we discuss results of [103–105] which relate critical percolation and ISE above the upper critical dimension, following the presentation of [188].

---

[4] In [188], convergence is proved as measures on probability measures on the one-point compactification of $\mathbb{R}^d$, but the one-point compactification can be avoided as in the proof of Theorem 16.6.

[5] The statement of Theorem 16.9 for $r = 3$ in [11, 12] incorrectly included the case where $t_j = 0$ for some j.

Consider independent Bernoulli bond percolation on $\mathbb{Z}^d$, nearest-neighbour or spread-out, with p fixed and equal to its critical value p_c. The basic definitions are laid out in Chap. 9. As usual, we let $C(0)$ denote the random set of vertices connected to 0, of cardinality $|C(0)|$. Let

$$\tau^{(2)}(x;n) = \mathbb{P}_{p_c}(C(0) \ni x, |C(0)| = n) \tag{16.62}$$

denote the probability at the critical point that the origin is connected to x via a cluster containing n vertices, and define the generating function

$$\tau_z^{(2)}(x) = \sum_{n=1}^{\infty} \tau^{(2)}(x;n)z^n, \tag{16.63}$$

which converges absolutely if $|z| \leq 1$. Assuming no infinite cluster at p_c, $\tau_1^{(2)}(x)$ is the probability that 0 is connected to x. Also, if we set $z = 1 - \gamma$, then $\hat{\tau}_z^{(2)}(0)$ is equal to $\chi(p, \gamma)$ of (9.29).

The definitions of the critical exponents η and δ in (9.17) and (9.18), together with (9.29), suggest that

$$\hat{\tau}_1^{(2)}(k) \sim \frac{c_1}{|k|^{2-\eta}} \text{ as } k \to 0, \quad \hat{\tau}_z^{(2)}(0) \sim \frac{c_2}{(1-z)^{1-1/\delta}} \text{ as } z \to 1^-. \tag{16.64}$$

Inserting the mean-field values $\eta = 0$ and $\delta = 2$, the simplest combination of the above asymptotic relations for $d > 6$ would be

$$\hat{\tau}_z^{(2)}(k) = \frac{C_2}{D_2^2|k|^2 + 2^{3/2}(1-z)^{1/2}} + \text{error}, \tag{16.65}$$

for some constants C_2, D_2. This is analogous to (16.8) with $\zeta = 1$. Theorem 11.3 shows that (16.65) is correct for sufficiently spread-out percolation above six dimensions, in the sense that there are functions $\epsilon_1(z)$ and $\epsilon_2(k)$ with $\lim_{z \to 1^-} \epsilon_1(z) = \lim_{k \to 0} \epsilon_2(k) = 0$, and constants C_2 and D_2 depending on d and L, such that

$$\hat{\tau}_z^{(2)}(k) = \frac{C_2}{D_2^2|k|^2 + 2^{3/2}(1-z)^{1/2}} [1 + \epsilon(z,k)] \tag{16.66}$$

for real $z \in [0, 1)$, with $|\epsilon(z,k)| \leq \epsilon_1(z) + \epsilon_2(k)$.

We have seen in (16.19) that the coefficient of z^n in $[|k|^2 + 2^{3/2}(1-z)^{1/2}]^{-1}$ is intimately related to $\hat{A}^{(2)}(k)$ and hence to ISE. Thus (16.66) is highly suggestive that ISE occurs as a scaling limit for percolation, but the control of the error term in (16.66) is too weak to obtain bounds on $\hat{\tau}^{(2)}(kD_2^{-1}n^{-1/4}; n)$ via contour integration. However, for the nearest-neighbour model in sufficiently high dimensions, better control of the error terms has been obtained, for *complex* z with $|z| < 1$, leading to the following theorem [103, 105]. The theorem also gives an asymptotic formula for the three-point function

$$\tau^{(3)}(x,y;n) = \mathbb{P}_{p_c}(x, y \in C(0), |C(0)| = n), \tag{16.67}$$

in terms of its Fourier transform

$$\hat{\tau}^{(3)}(k, l; n) = \sum_{x,y \in \mathbb{Z}^d} \tau^{(3)}(x, y; n) e^{ik \cdot x + il \cdot y}. \tag{16.68}$$

Theorem 16.10. *Let $p = p_c$. Fix $k, l \in \mathbb{R}^d$ and any $\epsilon \in (0, \frac{1}{2})$. There is a d_0 such that for nearest-neighbour percolation with $d \geq d_0$, there are constants C_2, D_2 (depending on d) such that as $n \to \infty$*

$$\hat{\tau}^{(2)}(kD_2^{-1}n^{-1/4}; n) = \frac{C_2}{\sqrt{8\pi n}} \hat{A}^{(2)}(k)[1 + O(n^{-\epsilon})], \tag{16.69}$$

$$\hat{\tau}^{(3)}(kD_2^{-1}n^{-1/4}, lD_2^{-1}n^{-1/4}; n) = \frac{C_2}{\sqrt{8\pi}} n^{1/2} \hat{A}^{(3)}(k+l, k, l)[1 + O(n^{-\epsilon})]. \tag{16.70}$$

It follows from (16.69) that

$$\begin{aligned}
\mathbb{P}_{p_c}(|C(0)| = n) &= n^{-1} \hat{\tau}^{(2)}(0; n) \\
&= C_2 (8\pi)^{-1/2} n^{-3/2}[1 + O(n^{-\epsilon})]. \tag{16.71}
\end{aligned}$$

Comparing with (9.19), this is a statement that the critical exponent δ is equal to 2 in high dimensions.

The proof of (16.70) considers the generating function

$$\hat{\tau}_z^{(3)}(k, l) = \sum_{n=1}^{\infty} \tau^{(3)}(k, l; n) z^n, \tag{16.72}$$

and shows that there is a positive constant V such that

$$\hat{\tau}_z^{(3)}(k, l) = V \hat{\tau}_z^{(2)}(k+l) \hat{\tau}_z^{(2)}(k) \hat{\tau}_z^{(2)}(l) + \text{error}, \tag{16.73}$$

where the error term is treated using Exercise 7.7. The variables in (16.70) are arranged schematically as:

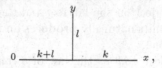

and (16.73) asserts that the three-point function approximately decouples into the vertex factor V times a product of three two-point functions. An asymptotic relation in the spirit of (16.73), with $k = l = 0$, was conjectured for $d > 6$ already in [12].

We expect that Theorem 16.10 should extend to general r-point functions, for all $r \geq 2$, but this has not been proven. This is essentially the conjecture of [103] that the scaling limit of the incipient infinite cluster is ISE for $d > 6$. We now discuss this conjecture in more detail.

Fix n and let S be a site lattice animal containing n sites, i.e., a set of n vertices in $\mathbb{Z}^d$ any two of which are connected by a sequence of vertices in S separated by unit distance. Supposing that S contains the origin, define the probability measure $\nu_S \in M_1(\mathbb{R}^d)$ to assign mass n^{-1} to $xD_2^{-1}n^{-1/4}$, for each $x \in S$. We define μ_n^{Perc} to be the probability measure on $M_1(\mathbb{R}^d)$ which assigns probability $\mathbb{P}_{p_c}(C(0) = S \,|\, |C(0)| = n)$ to ν_S, for each such S. We regard the limit of μ_n^{Perc}, as $n \to \infty$, as the scaling limit of the incipient infinite cluster. This is related to Kesten's definitions of the incipient infinite cluster [142] (see Sect. 11.3), but here we are conditioning the cluster to have a fixed large size rather than to reach a distant hyperplane, and, more importantly, here we are taking the lattice spacing to zero as $n \to \infty$. The conjecture of [103] is that, as in Corollary 16.8 above, μ_n^{Perc} converges weakly to μ_{ISE} for $d > 6$.

The conjecture is supported by Theorem 16.10. In fact, the characteristic functions $\hat{M}_n^{(1)}(k)$ and $\hat{M}_n^{(2)}(k,l)$ of the first and second moment measures $M_n^{(1)}$ and $M_n^{(2)}$ of μ_n^{Perc} are given by

$$\hat{M}_n^{(1)}(k) = \frac{\hat{\tau}^{(2)}(kD_2^{-1}n^{-1/4}; n)}{\hat{\tau}^{(2)}(0; n)}, \tag{16.74}$$

$$\hat{M}_n^{(2)}(k,l) = \frac{\hat{\tau}^{(3)}(kD_2^{-1}n^{-1/4}, lD_2^{-1}n^{-1/4}; n)}{\hat{\tau}^{(3)}(0, 0; n)}, \tag{16.75}$$

and in high dimensions these converge respectively to the characteristic functions $\hat{A}^{(2)}(k)$ and $\hat{A}^{(3)}(k+l, k, l)$ of the corresponding ISE moments, by Theorem 16.10 (see also (16.39)).

From this point of view, the IIC is regarded as a cluster in $\mathbb{R}^d$ arising in the scaling limit, represented by ISE. ISE is almost surely supported on a compact subset of $\mathbb{R}^d$, but on the scale of the lattice, this corresponds to an infinite cluster. The scaling here is consistent with the fact, discussed in Sect. 11.4.4, that in high dimensions the largest cluster in a box of radius r with bulk boundary condition has size of order r^4, and suggests that the largest cluster in a large box looks like an ISE cluster. In Sect. 17.3, a related but different construction is discussed for oriented percolation, where the IIC is first constructed as an infinite object on the lattice, and then the scaling limit of this is taken. This order of limits naturally produces an infinite continuous object, rather than a compact one.

ISE was defined in Sect. 16.3 as the law of a random probability measure on $\mathbb{R}^d$. However, ISE can also be formulated in such a way as to contain more detailed information including the structure of paths joining any finite set of points in the cluster, with these described by the functions $a^{(r)}(\vec{k}, \vec{t})$. See [152]. This is consistent with the approach of [4–6, 8] to the scaling limit, although here our focus is on a single percolation cluster, rather than on many clusters as in those references.

17

Super-Brownian Motion

Super-Brownian motion (SBM) is an active field in probability theory and has an extensive literature. Introductory accounts can be found in [152, 177], and a nontechnical introduction is [189]. SBM is a continuous time Markov process whose state space is the set of finite measures on $\mathbb{R}^d$. The subject can be technical, and some of the techniques used to study SBM require tools that many researchers interested in statistical mechanics do not have handy in their kit. Our account is introductory, and develops only those aspects that are essential for our specific purposes. In particular, attention is limited to the canonical measure of SBM.

In recent years, it has been realized that SBM is the scaling limit of several interacting particle systems at criticality. One aspect of this is the fact, already discussed in Sects. 16.4–16.5, that ISE describes the scaling limit of lattice trees in dimensions $d > 8$ and of critical percolation clusters in dimensions $d > 6$. In this chapter, we discuss results showing that the canonical measure of SBM describes the scaling limit of critical oriented percolation for $d + 1 > 4 + 1$, of the critical contact process for $d > 4$ (and also in lower dimensions when the infection range is suitably unbounded), and of lattice trees for $d > 8$. In addition, SBM is the scaling limit of the voter model in dimensions $d \geq 2$ [37, 58, 59], but the voter model will not be discussed further here. The relation between super-Brownian motion and the voter model, contact process, and oriented percolation is discussed in the introductory article [178].

17.1 The Canonical Measure of SBM

The canonical measure of SBM describes the scaling limit of critical branching random walk, started from a single particle at the origin as in the construction of ISE. With ISE, we conditioned critical branching random walk clusters to have exactly n particles, assigned mass n^{-1} to each particle, rescaled space by $n^{-1/4}$, and let $n \to \infty$. The result was a random probability measure on $\mathbb{R}^d$. For the canonical measure of SBM we will scale differently.

Let (T, φ) be a critical branching random walk configuration with law (15.8), as in Sect. 15.2. We make no assumption about the total number of vertices in T. Typically, T will have few vertices, but we are most interested in the rare trees that are long-lived. Let T_m denote the set of vertices at graph distance m from the root of T. We refer to T_m as the mth generation of T. Since T is almost surely finite, eventually $T_m = \varnothing$.

We introduce a scaling variable n, which is a large integer that will be sent to infinity. Four simultaneous scalings are made, as follows.

- *Scaling of time.* We define a time variable $t = \frac{m}{n}$. This fixes attention on the individuals T_m in generations m of order n.
- *Scaling of space.* In a configuration (T, φ), an individual in generation n (if T lives so long) is mapped by φ to the endpoint of an n-step random walk path. Such a point is typically at distance order $n^{1/2}$ from the origin, so we rescale the lattice to $n^{-1/2}\mathbb{Z}^d$.
- *Scaling of mass.* A critical tree that survives to a generation of order n typically has order n members in that generation. To obtain a generation mass of order 1, we assign mass n^{-1} to each embedded vertex.
- *Scaling of probability.* By a theorem of Kolmogorov, the probability that a critical tree survives for order n generations is of order n^{-1}. The long-lived trees of interest thus have a vanishingly small probability. To compensate, we multiply probabilities by n, so that survival to order n generations will have measure of order 1, rather than probability of order n^{-1}. This produces an unnormalized measure, rather than a probability measure.

The above is carried out, in detail, as follows. Given a critical branching random walk configuration (T, φ), we rescale the lattice to $n^{-1/2}\mathbb{Z}^d$, and denote by $R_n^{(m/n)}$ the mass distribution in $n^{-1/2}\mathbb{Z}^d$ of the mth generation T_m of T. Explicitly, $R_n^{(m/n)}$ is the discrete finite measure on $\mathbb{R}^d$ that places mass n^{-1} at each embedded vertex of T_m, with multiplicity, i.e.,

$$R_n^{(m/n)} = \frac{1}{n} \sum_{i \in T : |i| = m} \delta_{n^{-1/2}\varphi(i)}. \qquad (17.1)$$

Note that $R_n^{(m/n)}$ need not be a probability measure. The measure $R_n^{(m/n)}$ is a *random* measure. Let $\mathcal{R}_n^{(m/n)}$ denote the probability law of this random measure. Thus $\mathcal{R}_n^{(m/n)}$ is a probability measure on the space $M_F(\mathbb{R}^d)$ of finite measures on $\mathbb{R}^d$, which quantifies how likely it is that a particular mass distribution occurs for the mth generation of a critical branching random walk in $n^{-1/2}\mathbb{Z}^d$. So far, we have rescaled generation m to time m/n, space to $n^{-1/2}\mathbb{Z}^d$, and vertex mass has been set equal to n^{-1}.

It remains to scale the probability measure $\mathcal{R}_n^{(m/n)}$. When m is of order n, the random measure $R_n^{(m/n)}$ is the zero measure on $\mathbb{R}^d$ with probability $1 - O(n^{-1})$, since the probability of survival for m generations is order $m^{-1} \approx n^{-1}$. To amplify the rare event of survival to m generations, we consider the measure $n\mathcal{R}_n^{(m/n)}$ on $M_F(\mathbb{R}^d)$, which has total measure n. This measure places measure $n - O(1)$ on the zero measure on $\mathbb{R}^d$, and the remaining measure $O(1)$ on the nontrivial mass distributions due to the rare trees that survive for m generations.

It is known that the limit of $n\mathcal{R}_n^{(\lfloor tn \rfloor/n)}$, as $n \to \infty$, consists of an infinite point mass on the zero measure on $\mathbb{R}^d$, plus a nontrivial finite measure $\mathcal{R}^{(t)}$ on $M_0(\mathbb{R}^d)$, the finite measures on $\mathbb{R}^d$ excluding the zero measure (see [177]). More precisely, there is a measure $\mathcal{R}^{(t)}$ on $M_0(\mathbb{R}^d)$ such that $n\mathcal{R}_n^{(\lfloor tn \rfloor/n)}$ converges weakly to $\mathcal{R}^{(t)}$ plus an infinite point mass on the zero measure on $\mathbb{R}^d$. In particular, for every bounded continuous function f on $M_0(\mathbb{R}^d)$,

$$\lim_{n\to\infty} n \int_{M_0(\mathbb{R}^d)} f(R) \mathrm{d}\mathcal{R}_n^{(\lfloor tn \rfloor/n)}(R) = \int_{M_0(\mathbb{R}^d)} f(R)\mathrm{d}\mathcal{R}^{(t)}(R). \qquad (17.2)$$

In other words, when normalized by n, the average of a function over the long-lived, random configurations of the discrete mass distribution defined by $\mathcal{R}_n^{(\lfloor tn \rfloor/n)}$ converges to a corresponding integral over the configurations of the continuous mass distribution defined by $\mathcal{R}^{(t)}$.

Moreover, stronger results [178, Chap. 7] assert the convergence of the scaling limit in the sense of convergence of measures on functions $R^{(\cdot)}$ from the time interval $(0, \infty)$ into $M_F(\mathbb{R}^d)$. The measure-valued path $R^{(\cdot)}$ represents the evolving mass distribution of the embedded critical tree. The *canonical measure of super-Brownian motion*, denoted $\mathbb{N}_0$, is a measure on measure-valued paths that gives the appropriate weight to $R^{(\cdot)}$. Fig. 17.1 illustrates the evolution.

17.2 Moment Measures of the Canonical Measure

Later we will state results showing convergence of critical oriented percolation and the critical contact process to the canonical measure, in the sense of convergence of the moment measures. This is a form of convergence of finite-dimensional distributions. The results make use of explicit formulas for the moment measures of the canonical measure of SBM, and we now discuss these formulas.

By definition, the lth moment measure of the canonical measure $\mathbb{N}_0$ has Fourier transform

$$\hat{M}_t^{(l)}(\bar{k}) = \mathbb{N}_0\Big(\int_{\mathbb{R}^{dl}} \mathrm{d}R^{(t_1)}(x_1) \cdots \mathrm{d}R^{(t_l)}(x_l) \prod_{j=1}^{l} e^{ik_j \cdot x_j} \Big). \qquad (17.3)$$

Here $\bar{k} = (k_1, \ldots, k_l)$ with each $k_j \in \mathbb{R}^d$, and $R^{(\cdot)}$ has distribution $\mathbb{N}_0$, so that each $R^{(t)}$ is a non-negative finite measure on $\mathbb{R}^d$. It is possible to guess the form of (17.3) by thinking about the approximating branching random walk. In this section, we explain the guess, and sketch a proof for the case of critical Poisson branching considered in Sect. 15.2.

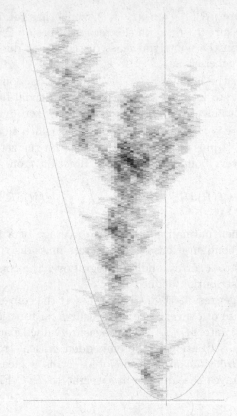

Fig. 17.1. The evolution of a mass distribution under the canonical measure, with time vertical and 1-dimensional space horizontal. Darker shading represents higher mass, and the parabola $t = x^2$ shows the spatial scaling.

According to the description of the canonical measure in Sect. 17.1, the discrete analogue of (17.3) is

$$\hat{M}_{n,\bar{t}}^{(l)}(\bar{k}) = \frac{1}{n^{l-1}} \hat{S}_{\lfloor n\bar{t} \rfloor}^{(l+1)}(\bar{k}/\sqrt{n}), \qquad (17.4)$$

where

$$S_{\bar{m}}^{(l+1)}(\bar{x}) = \sum_{(T,\varphi)} \mathbb{P}(T,\varphi) \sum_{\substack{i_1,\ldots,i_l \in T \\ |i_j| = m_j \, \forall j}} \prod_{j=1}^{l} I[\varphi(i_j) = x_j]. \qquad (17.5)$$

The factors involving n in (17.4) arise as follows. The scaling of $\bar{k}$ by $n^{-1/2}$ corresponds to scaling the lattice spacing, the scaling of $\bar{t}$ by n is the temporal scaling, a factor n^{-l} is due to mass n^{-1} at each of the l vertices $x_1,\ldots,x_l$, and an additional factor n changes n^{-l} to $n^{-(l-1)}$ to rescale probability.

Recall the notions of shape, skeleton and compatibility from Sect. 15.2. Suppose that $(\sigma; \vec{y}, \vec{s})$ is compatible with $(T, \varphi, \bar{\imath})$ contributing to the sum in (17.5), where $\sigma \in \Sigma_{l+1}$, and where, as in Sect. 15.2, we write e.g. $\vec{s} = (s_1, \ldots, s_{2l-1})$ for $(2l-1)$-component vectors (l corresponds to $r-1$). Compatibility forces relationships between $\vec{s}$ and $\bar{m}$ and between $\vec{y}$ and $\bar{x}$. To be specific about these relationships, we make the following definitions.

Given a shape $\sigma \in \Sigma_{l+1}$, for each vertex j of degree 1 in σ, other than vertex 0, we let ω_j be the set of edges in σ on the path from 0 to j ($j = 1, \ldots, l$). Given $\vec{s} \in \mathbb{R}_+^{2l-1}$, we define the jth component $\bar{m}_j(\sigma, \vec{s})$ of $\bar{m}(\sigma, \vec{s}) \in \mathbb{R}_+^l$ by

$$\bar{m}_j(\sigma, \vec{s}) = \sum_{i \in \omega_j} s_i, \qquad (17.6)$$

and compatibility requires (17.6). For $t_i > 0$ we also define an $(l-1)$-dimensional subset $J_{\bar{t}}(\sigma)$ of $\mathbb{R}_+^{2l-1}$ by

$$J_{\bar{t}}(\sigma) = \{\vec{s} : \bar{m}(\sigma, \vec{s}) = \bar{t}\}. \qquad (17.7)$$

For example, for $l = 2$, there is a unique shape σ and we have simply

$$J_{\bar{t}}(\sigma) = \{(s, t_1 - s, t_2 - s) : s \in [0, t_1 \wedge t_2]\}. \qquad (17.8)$$

The relation between $\bar{m}$ and $\vec{s}$ has a natural counterpart for $\bar{x}$ and $\vec{y}$. This translates into a dual relation for the Fourier variable $\bar{k}$, which we recall from Sect. 16.3. Given a shape $\sigma \in \Sigma_{l+1}$ and $\bar{k} = (k_1 \ldots, k_l) \in \mathbb{R}^{dl}$, we define $\vec{k}(\sigma) \in \mathbb{R}^{(2l-1)d}$ by setting the ith component $\vec{k}_i(\sigma) \in \mathbb{R}^d$ ($i = 1, \ldots, 2l-1$) of $\vec{k}(\sigma)$ to be

$$\vec{k}_i(\sigma) = \sum_{j=1}^{l} k_j I[i \in \omega_j], \qquad (17.9)$$

where, on the right hand side, k_j denotes the jth component of $\bar{k}$ and I is an indicator function.

For $l \geq 3$, there are redundancies in replacing the sum over $(T, \varphi, \bar{\imath})$ in $S_{\bar{m}}^{(l+1)}(\bar{x})$ by an appropriate sum over $(\sigma; \vec{y}, \vec{s})$, as discussed in the proof of Theorem 16.5. However, such overcounting arises from degenerate configurations with at least one s_j equal to zero, and it can be shown that these are lower order and do not contribute to the limit. Therefore, in place of the right hand side of (17.4), we consider instead

$$\sum_{\sigma \in \Sigma_{l+1}} \frac{1}{n^{l-1}} \sum_{\vec{s} \in J_{\lfloor n\bar{t} \rfloor}(\sigma) \cap \mathbb{Z}_+^{l-1}} \sum_{N=0}^{\infty} \hat{b}_N^{(l+1)}(\sigma; \vec{k}(\sigma)n^{-1/2}, \vec{s}) \qquad (17.10)$$

(where $\hat{b}_N$ is the Fourier transform of (15.23) with respect to $\vec{y}$), which can be shown to have the same limit as (17.4) when $n \to \infty$. By Theorem 15.4, the quantity $\sum_{N=0}^{\infty} b_N^{(l+1)}(\sigma; \vec{k}n^{-1/2}, \vec{s})$ is the coefficient of $\prod_{j=1}^{2l-1} \zeta_j^{s_j}$ in the generating function

$$\prod_{j=1}^{2l-1} \frac{1}{1 - \zeta_j \hat{D}(k_j n^{-1/2})}. \tag{17.11}$$

As in (16.18), this coefficient converges to a product of Gaussians. The set $J_{\lfloor n\bar{t} \rfloor}(\sigma)$ is $(l-1)$-dimensional, and the factor $n^{-(l-1)}$ converts the sum into a Riemann sum. It is therefore entirely plausible, and it can be proven, that if $t_i > 0$ for all i then

$$\lim_{n \to \infty} \hat{M}_{n,\bar{t}}^{(l)}(\vec{k}) = \begin{cases} e^{-|k|^2 t/2d} & (l = 1), \\ \sum_{\sigma \in \Sigma_{l+1}} \int_{J_{\bar{t}}(\sigma)} d\vec{s} \prod_{j=1}^{2l-1} e^{-|k_j(\sigma)|^2 s_j/2d} & (l \geq 2). \end{cases} \tag{17.12}$$

A different but related and more general proof of (17.12) is discussed in [109]. The right hand side of (17.12) is equal to $\hat{M}_{\bar{t}}^{(l)}(\vec{k})$ of (17.3). The latter fact is essentially [2, Theorem 3.1] (see also [69] and [152, Proposition IV.2.(ii)]).

Exercise 17.1. Fill in the details missing in the above discussion to give a complete proof of (17.12).

Explicitly, for $l = 2$,

$$\hat{M}_{t_1,t_2}^{(2)}(k_1, k_2) = \int_0^{t_1 \wedge t_2} ds\, e^{-|k_1+k_2|^2 s/2d} e^{-|k_1|^2(t_1-s)/2d} e^{-|k_2|^2(t_2-s)/2d}. \tag{17.13}$$

In x-space, recalling the definition of the Gaussian density $p_t(x)$ in (6.8), this is equivalent to the density

$$M_{t_1,t_2}^{(2)}(x_1, x_2) = \int_0^{t_1 \wedge t_2} ds \int_{\mathbb{R}^d} dy\, p_s(y) p_{t_1-s}(x_1 - y) p_{t_2-s}(x_2 - y). \tag{17.14}$$

This last formula clearly shows the branching structure. Mass arrives at x_1 at time t_1 and at x_2 at time t_2 via a Brownian path that leaves the origin and splits into two at a time s chosen uniformly from $[0, t_1 \wedge t_2]$, with the two particles then continuing to x_1 and x_2 at their appointed times.

Exercise 17.2. Starting with the right hand side of (17.12) as an expression for the Fourier transform $\hat{M}_{\bar{t}}^{(l)}(\vec{k})$ of the moment measures, prove the recursion relation

$$\hat{M}_{\bar{t}}^{(l)}(\vec{k}) = \int_0^t dt\, \hat{M}_t^{(1)}(k_1 + \cdots + k_l) \sum_{I \subset J_1 : |I| \geq 1} \hat{M}_{\bar{t}_I - t}^{(i)}(\vec{k}_I) \hat{M}_{\bar{t}_{J \backslash I} - t}^{(l-i)}(\vec{k}_{J \backslash I}) \tag{17.15}$$

for $l \geq 2$, where $i = |I|$, $J = \{1, \ldots, l\}$, $J_1 = J \backslash \{1\}$, $\underline{t} = \min_i t_i$, $\bar{t}_I$ denotes the vector consisting of the components t_i of $\bar{t}$ with $i \in I$, and $\bar{t}_I - t$ denotes subtraction of t from each component of $\bar{t}_I$.

A solution to a slightly weakened version of the following exercise can be found in [114].

Exercise 17.3. Prove that for $l \geq 1$, $t_1 \geq s = \max\{t_2,\ldots,t_l\}$ and $k_j \in \mathbb{R}^d$,

$$\hat{M}^{(l)}_{t_1,t_2,\ldots,t_l}(0,k_2,\ldots,k_l) = \hat{M}^{(l)}_{s,t_2,\ldots,t_l}(0,k_2,\ldots,k_l). \tag{17.16}$$

A solution to the following exercise can be found in [114].

Exercise 17.4. Prove that for $l \geq 1$,

$$\hat{M}^{(l)}_{t,\ldots,t}(\vec{0}) = t^{l-1}2^{-l+1}l!. \tag{17.17}$$

17.3 Critical Oriented Percolation and SBM

For $r \geq 2$, the oriented percolation r-point functions are defined by

$$\tau^{(r)}_{p,\vec{n}}(\vec{x}) = \mathbb{P}_p((0,0) \to (x_i,n_i) \text{ for each } i = 1,\ldots,r-1). \tag{17.18}$$

The event on the right hand side makes no statement about the occurrence of $(x_i,n_i) \to (x_j,n_j)$ for any $i \neq j$. The following theorem is proved in [121]. In its statement, the constants A and v are the same as those appearing in Theorem 12.3.

Theorem 17.5. *Consider the spread-out model of oriented percolation. Let $d > 4$, $p = p_c$, $\delta \in (0,1 \wedge \frac{d-4}{2})$, $r \geq 2$, $\bar{t} = (t_1,\ldots,t_{r-1}) \in (0,\infty)^{r-1}$, and $\bar{k} = (k_1,\ldots,k_{r-1}) \in \mathbb{R}^{(r-1)d}$. There is a constant $V = V(d,L)$, with $|V-1| \leq CL^{-d}$, and an $L_0 = L_0(d)$ (independent of r) such that for $L \geq L_0$,*

$$\hat{\tau}^{(r)}_{p_c,\lfloor n\bar{t}\rfloor}(\bar{k}/\sqrt{vn}) = n^{r-2}V^{r-2}A^{2r-3}\left[\hat{M}^{(r-1)}_{\bar{t}}(\bar{k}) + O(n^{-\delta})\right], \tag{17.19}$$

with the error estimate uniform in $\bar{k}$ in a bounded subset of $\mathbb{R}^{(r-1)d}$ (but not in L or the t_j).

The theorem is proved using induction on r. The induction is started at $r = 2$ with Theorem 12.3, which gives control of the critical two-point function (and which was proved using induction on n). The induction on r is advanced using a second expansion for the r-point functions for $r \geq 3$, which goes beyond the expansion methods of Chap. 13.

For $r = 3$, the main term on the right hand side of (17.19) reduces to

$$nVA^3 \int_0^{t_1 \wedge t_2} ds\, e^{-|k_1+k_2|^2 s/2d}e^{-|k_1|^2(t_1-s)/2d}e^{-|k_2|^2(t_2-s)/2d}. \tag{17.20}$$

Equation (17.20) can be interpreted as indicating that a cluster connecting the origin to $(x_1,\lfloor nt_1\rfloor)$ and $(x_2,\lfloor nt_2\rfloor)$, with the x_i of order $n^{1/2}$, can be considered to decompose into a product of three independent two-point functions joined together at a branch point. Each two-point function gives rise to a Gaussian, together with a factor A, according to Theorem 12.3(a). This decomposition into independent two-point functions is not exact, but is compensated by the vertex factor V associated with the branch point. The integral with respect to s corresponds to a sum over possible temporal locations of the branch point, with the additional factor n accounting for the change from a sum to an integral.

Similar considerations apply to (17.19), with additional structure due to the proliferation of shapes. There are $r - 2$ branch points in the general case, each contributing nV, and $2r - 3$ two-point functions, each contributing A times a Gaussian. The integral over $J_{\bar{t}}(\sigma)$ corresponds to a sum over time intervals between the various branch points, and is constrained so that the shape's leaves are specified by the times $\lfloor n\bar{t} \rfloor$.

The tree graph bounds (derivations in [12, 85] extend easily to the oriented setting), together with the bound on the two-point function of Theorem 15.2(a), immediately imply that the left hand side of (17.19) is bounded above by a multiple of n^{r-2}. By (17.19), this elementary upper bound gives the correct power of n, above the upper critical dimension.

Theorem 17.5 can be rephrased to say that, under its hypotheses, the moment measures of rescaled critical oriented percolation converge to those of the canonical measure of super-Brownian motion. We now make this interpretation more explicit.

First, for $t \in (0, \infty)$, we define $X_{n,t}$ to be the discrete finite random measure on $\mathbb{R}^d$ placing mass $(A^2 V n)^{-1}$ at each vertex at time $\lfloor nt \rfloor$ in $(vn)^{-1/2} C(0,0)$. The characteristic function of the lth moment measure of the measure-valued process $X_{n,\cdot}$ is defined as in (17.3) and is given by

$$\hat{N}_{n,\bar{t}}^{(l)}(\bar{k}) = (A^2 V n)^{-l} \hat{\tau}_{\lfloor n\bar{t} \rfloor}^{(l+1)}(\bar{k}/\sqrt{vn}). \qquad (17.21)$$

It then follows immediately from Theorems 15.2 and 17.5 that, under their hypotheses, for all $l \geq 1$,

$$\lim_{n \to \infty} AVn \hat{N}_{n,\bar{t}}^{(l)}(\bar{k}) = \hat{M}_{\bar{t}}^{(l)}(\bar{k}). \qquad (17.22)$$

This can be regarded as a statement that $AVn\mathbb{P}(\{X_{n,t}\}_{t>0} \in \cdot)$ converges to $\mathbb{N}_0$ in the sense of convergence of finite-dimensional distributions.

This shows that spread-out critical oriented percolation and critical branching random walk have the same scaling limit, for $d > 4$ (compare [177, Theorem II.7.3(a)]). A crucial difference between oriented percolation and branching random walk is that particles can coexist at the same vertex for the latter, but not for the former.

The above gives a statement of convergence of finite-dimensional distributions. To prove weak convergence, as a measure-valued process, of rescaled spread-out oriented percolation for $d > 4$ to the canonical measure of super-Brownian motion, it would be necessary also to prove tightness. Currently, this is an open problem.

By Theorem 12.7, together with (12.38) which asserts that the survival probability is asymptotic to $(Bn)^{-1}$, we have that $B = AV/2$. This in turn implies that the factor AVn in (17.22) corresponds asymptotically to twice the reciprocal of the survival probability.

The incipient infinite cluster of spread-out oriented percolation in dimensions $d + 1 > 4 + 1$ was discussed in Sect. 12.4. In [121], the natural conjecture was formulated that the scaling limit of the incipient infinite cluster for oriented percolation in dimensions $d + 1 > 4 + 1$ is the canonical measure of super-Brownian motion, conditioned on survival for all time. The proof makes use of the IIC r-point functions, defined by

$$\rho_{\bar{m}}^{(r)}(\bar{y}) = \mathbb{P}_\infty((0,0) \to (y_i, m_i) \text{ for each } i = 1, \dots, r - 1), \qquad (17.23)$$

where $\mathbb{P}_\infty$ is the measure of Theorem 12.5. It is proved in [114] that the scaling limit of the IIC r-point functions is given by

$$\lim_{m \to \infty} \frac{1}{(mA^2V)^{r-1}} \hat{\rho}_{\lfloor m\bar{t} \rfloor}^{(r)}(\bar{k}/\sqrt{vm}) = \hat{M}_{1,\bar{t}}^{(r)}(0, \bar{k}), \qquad (17.24)$$

for all $r \geq 2$, $\bar{t} = (t_1, \dots, t_{r-1}) \in (0,1]^{r-1}$ and $\bar{k} \in \mathbb{R}^{d(r-1)}$, and it is pointed out in [109] that this proves the conjecture at the level of convergence of finite-dimensional distributions.

Solutions to the following two exercises can be found in [114]. Each makes use of the following extension of Theorem 17.5 that was proved in [121]. Let m denote the second largest component of $\bar{n} = (n_1, \dots, n_{r-1})$. Then [121, (2.52)] states that for each $r \geq 3$,

$$\hat{\tau}_{\bar{n}}^{(r)}(\bar{k}/\sqrt{v\,n}) = n^{r-2}V^{r-2}A^{2r-3}\left[\hat{M}_{\bar{n}/n}^{(r-1)}(\bar{k}) + \mathcal{O}((m+1)^{-\delta})\right] \qquad (17.25)$$

holds uniformly in $n \geq m$.

Exercise 17.6. Prove Theorem 12.7(a), using (17.25), Exercises 17.3 and 17.4, and

$$\mathbb{E}_\infty[N_m^l] = \lim_{n \to \infty} \frac{\hat{\tau}_{n,m,\dots,m}^{(l+2)}(0)}{\hat{\tau}_n^{(2)}(0)}. \qquad (17.26)$$

Exercise 17.7. Prove (17.24) for all $r \geq 2$. Use Exercise 17.3 and (17.25).

17.4 The Critical Contact Process and SBM

A version of Theorem 17.5 is proven in [117] for the contact process. This shows that in dimensions $d > 4$ the spread-out contact process with L sufficiently large converges to the canonical measure of super-Brownian motion in the sense of convergence of finite-dimensional distributions. The proof uses the approximation by oriented percolation discussed in Sect. 14.2. In addition, a similar result is obtained in [117] for dimensions $1 \le d \le 4$, for the model with $L = L_1 T^b$ with parameters as in Theorem 14.3. In particular, the assumption on the exponent b is that $b > \frac{4-d}{2d}$, as in (14.24).

This should be compared with the results of [68], who proved convergence to super-Brownian motion for the model of the contact process with L proportional to T for $d > 2$ and proportional to $(T \log T)^{1/2}$ for $d = 2$. This requires faster growth in T for $d > 2$ but is sharper for $d = 2$. The results of [68] use completely different methods than the lace expansion methods discussed in these lecture notes. In particular, they work entirely with continuous time and do not use an approximation by oriented percolation. The convergence statements of [68] are stronger than those of [117] in the sense that they also prove tightness and thus go beyond the convergence of moment measures. In addition, their results are not formulated in terms of the canonical measure, but rather prove convergence to super-Brownian motion with an arbitrary finite measure as initial condition. On the other hand, their assumption that $b = 1$ in dimensions $d > 4$ is considerably stronger than the assumption in [117] that L is large but finite for dimensions $d > 4$.

17.5 Lattice Trees and SBM

Corollary 16.8 shows that in dimensions $d > 8$ spread-out lattice trees converge to ISE when appropriately rescaled. In that scaling, we considered lattice trees with exactly n bonds, mass $(n + 1)^{-1}$ was placed at each vertex, space was rescaled by a multiple of $n^{-1/4}$, and the limit $n \to \infty$ was taken. Now we consider instead the scaling of Sect. 17.1, appropriate to give the canonical measure of SBM as limit.

For this, we need a time variable for lattice trees. The time variable is graph distance from the origin, i.e., vertices at "time" m are those at graph distance m from the origin in the lattice tree. Let $|x|_g$ denote the graph distance of x, and for $r \ge 2$, $\bar{m} = (m_1, \ldots, m_{r-1})$, and $\bar{x} = (x_1, \ldots, x_{r-1})$ such that $|x_i|_g = m_i$ for all i, define

$$t_{\bar{m}}^{(r)}(\bar{x}) = \sum_{T:T \ni 0, x_1, \ldots, x_{r-1}} z_c^{|T|}. \tag{17.27}$$

Note that the activity z dual to the size of the lattice tree has been set equal to its critical value z_c, and that lattice trees of arbitrary size contribute to (17.27).

We then rescale $\bar{x}$ by $n^{-1/2}$ and rescale $\bar{m}$ by n^{-1}. Under this scaling, the analogue of (17.19) is proved in [125] to hold for sufficiently spread-out lattice trees in dimensions $d > 8$. The proof uses an adaptation of the induction method of [120] for the two-point function, and then applies induction on r for the r-point functions with $r \ge 3$. The induction on r is advanced using the lace expansion on a tree [124] in place of the more difficult double expansion used in [62, 99].

References

1. J. Adler, Y. Meir, A. Aharony, and A.B. Harris. Series study of percolation moments in general dimension. *Phys. Rev. B*, **41**:9183–9206, (1990).
2. R.J. Adler. Superprocess local and intersection local times and their corresponding particle pictures. In E. Çinlar, K.L. Chung, and M.J. Sharpe, editors, *Seminar on Stochastic Processes 1992*, pages 1–42, Boston, (1993). Birkhäuser.
3. M. Aizenman. Geometric analysis of φ^4 fields and Ising models, Parts I and II. *Commun. Math. Phys.*, **86**:1–48, (1982).
4. M. Aizenman. The geometry of critical percolation and conformal invariance. In Hao Bai-Lin, editor, *STATPHYS 19, Proceedings Xiamen 1995*. World Scientific, (1996).
5. M. Aizenman. On the number of incipient spanning clusters. *Nucl. Phys. B [FS]*, **485**:551–582, (1997).
6. M. Aizenman. Scaling limit for the incipient spanning clusters. In *Mathematics of multiscale materials*, pages 1–24, IMA Vol. Math. Appl., 99. Springer, New York, (1998).
7. M. Aizenman and D.J. Barsky. Sharpness of the phase transition in percolation models. *Commun. Math. Phys.*, **108**:489–526, (1987).
8. M. Aizenman and A. Burchard. Hölder regularity and dimension bounds for random curves. *Duke Math. J.*, **99**:419–453, (1999).
9. M. Aizenman and R. Fernández. On the critical behavior of the magnetization in high dimensional Ising models. *J. Stat. Phys.*, **44**:393–454, (1986).
10. M. Aizenman and R. Fernández. Critical exponents for long-range interactions. *Lett. Math. Phys.*, **16**:39–49, (1988).
11. M. Aizenman, H. Kesten and C.M. Newman. Uniqueness of the infinite cluster and continuity of connectivity functions for short and long range percolation. *Commun. Math. Phys.*, **111**:505–531, (1987).
12. M. Aizenman and C.M. Newman. Tree graph inequalities and critical behavior in percolation models. *J. Stat. Phys.*, **36**:107–143, (1984).
13. M. Ajtai, J. Komlós, and E. Szemerédi. Largest random component of a k-cube. *Combinatorica*, **2**:1–7, (1982).
14. D. Aldous. The continuum random tree III. *Ann. Probab.*, **21**:248–289, (1993).
15. D. Aldous. Tree-based models for random distribution of mass. *J. Stat. Phys.*, **73**:625–641, (1993).

16. S.E. Alm and S. Janson. Random self-avoiding walks on one-dimensional lattices. *Commun. Statist.–Stochastic Models*, **6**:169–212, (1990).

17. N. Alon, I. Benjamini, and A. Stacey. Percolation on finite graphs and isoperimetric inequalities. *Ann. Probab.*, **32**:1727–1745, (2004).

18. M.N. Barber and B.W. Ninham. *Random and Restricted Walks: Theory and Applications.* Gordon and Breach, New York, (1970).

19. M.T. Barlow and E.A. Perkins. On the filtration of historical Brownian motion. *Ann. Probab.*, **22**:1273–1294, (1994).

20. D.J. Barsky and M. Aizenman. Percolation critical exponents under the triangle condition. *Ann. Probab.*, **19**:1520–1536, (1991).

21. D.J. Barsky and C.C. Wu. Critical exponents for the contact process under the triangle condition. *J. Stat. Phys.*, **91**:95–124, (1998).

22. C. Bezuidenhout and G. Grimmett. The critical contact process dies out. *Ann. Probab.*, **18**:1462–1482, (1990).

23. C. Bezuidenhout and G. Grimmett. Exponential decay for subcritical contact and percolation processes. *Ann. Probab.*, **19**:984–1009, (1991).

24. P. Billingsley. *Convergence of Probability Measures.* John Wiley and Sons, New York, (1968).

25. B. Bollobás. The evolution of random graphs. *Trans. Amer. Math. Soc.*, **286**:257–274, (1984).

26. B. Bollobás. *Random Graphs.* Cambridge University Press, Cambridge, 2nd edition, (2001).

27. B. Bollobás and Y. Kohayakawa. Percolation in high dimensions. *Europ. J. Combinatorics*, **15**:113–125, (1994).

28. B. Bollobás, Y. Kohayakawa, and T. Łuczak. The evolution of random subgraphs of the cube. *Random Struct. Alg.*, **3**:55–90, (1992).

29. E. Bolthausen and C. Ritzmann. A central limit theorem for convolution equations and weakly self-avoiding walks. To appear in *Ann. Probab.*

30. C. Borgs, J.T. Chayes, R. van der Hofstad, and G. Slade. Mean-field lattice trees. *Ann. Combinatorics*, **3**:205–221, (1999).

31. C. Borgs, J.T. Chayes, R. van der Hofstad, G. Slade, and J. Spencer. Random subgraphs of finite graphs: I. The scaling window under the triangle condition. *Random Struct. Alg.*, **27**:137–184, (2005).

32. C. Borgs, J.T. Chayes, R. van der Hofstad, G. Slade, and J. Spencer. Random subgraphs of finite graphs: II. The lace expansion and the triangle condition. *Ann. Probab.*, **33**:1886–1944, (2005).

33. C. Borgs, J.T. Chayes, R. van der Hofstad, G. Slade, and J. Spencer. Random subgraphs of finite graphs: III. The phase transition for the n-cube. To appear in *Combinatorica*.

34. C. Borgs, J.T. Chayes, H. Kesten, and J. Spencer. Uniform boundedness of critical crossing probabilities implies hyperscaling. *Random Struct. Alg.*, **15**:368–413, (1999).

35. C. Borgs, J.T. Chayes, H. Kesten, and J. Spencer. The birth of the infinite cluster: finite-size scaling in percolation. *Commun. Math. Phys.*, **224**:153–204, (2001).

36. A. Bovier, J. Fröhlich, and U. Glaus. Branched polymers and dimensional reduction. In K. Osterwalder and R. Stora, editors, *Critical Phenomena, Random Systems, Gauge Theories*, Amsterdam, (1986). North-Holland.

37. M. Bramson, J.T. Cox, and J-F. Le Gall. Super-Brownian limits of voter model clusters. *Ann. Probab.*, **29**:1001–1032, (2001).

38. E. Brézin, J.C. Le Guillou, and J. Zinn-Justin. Approach to scaling in renormalized perturbation theory. *Phys. Rev. D*, **8**:2418–2430, (1973).

39. D. Brydges, S.N. Evans, and J.Z. Imbrie. Self-avoiding walk on a hierarchical lattice in four dimensions. *Ann. Probab.*, **20**:82–124, (1992).

40. D.C. Brydges. A short course on cluster expansions. In K. Osterwalder and R. Stora, editors, *Critical Phenomena, Random Systems, Gauge Theories*, Amsterdam, (1986). North-Holland. Les Houches 1984.

41. D.C. Brydges and J.Z. Imbrie. Branched polymers and dimensional reduction. *Ann. Math.*, **158**:1019–1039, (2003).

42. D.C. Brydges and J.Z. Imbrie. Dimensional reduction formulas for branched polymer correlation functions. *J. Stat. Phys.*, **110**:503–518, (2003).

43. D.C. Brydges and J.Z. Imbrie. End to end distance from the Green's function for a hierarchical self-avoiding walk in four dimensions. *Commun. Math. Phys.*, **239**:523–547, (2003).

44. D.C. Brydges and J.Z. Imbrie. Green's function for a hierarchical self-avoiding walk in four dimensions. *Commun. Math. Phys.*, **239**:549–584, (2003).

45. D.C. Brydges and T. Spencer. Self-avoiding walk in 5 or more dimensions. *Commun. Math. Phys.*, **97**:125–148, (1985).

46. R.M. Burton and M. Keane. Density and uniqueness in percolation. *Commun. Math. Phys.*, **121**:501–505, (1989).

47. S. Caracciolo, G. Parisi, and A. Pelissetto. Random walks with short-range interaction and mean-field behavior. *J. Stat. Phys.*, **77**:519–543, (1994).

48. J.T. Chayes and L. Chayes. Ornstein-Zernike behavior for self-avoiding walks at all noncritical temperatures. *Commun. Math. Phys.*, **105**:221–238, (1986).

49. J.T. Chayes and L. Chayes. Percolation and random media. In K. Osterwalder and R. Stora, editors, *Critical Phenomena, Random Systems, Gauge Theories*, Amsterdam, (1986). North-Holland. Les Houches 1984.

50. J.T. Chayes and L. Chayes. On the upper critical dimension of Bernoulli percolation. *Commun. Math. Phys.*, **113**:27–48, (1987).

51. J.T. Chayes, L. Chayes, and R. Durrett. Inhomogeneous percolation problems and incipient infinite clusters. *J. Phys. A: Math. Gen.*, **20**:1521–1530, (1987).

52. Y. Cheng. *Long range self-avoiding random walks above critical dimension.* PhD thesis, Temple University, (2000).

53. N. Clisby, R. Liang and G. Slade. Self-avoiding walk enumeration via the lace expansion. In preparation.

54. A.R. Conway, I.G. Enting, and A.J. Guttmann. Algebraic techniques for enumerating self-avoiding walks on the square lattice. *J. Phys. A: Math. Gen.*, **26**:1519 1534, (1993).

55. A.R. Conway and A.J. Guttmann. Lower bound on the connective constant for square lattice self-avoiding walks. *J. Phys. A: Math. Gen.*, **26**:3719–3724, (1993).

56. R.M. Corless, G.H. Gonnet, D.E.G. Hare, D.J. Jeffrey, and D.E. Knuth. On the Lambert W function. *Adv. Comput. Math.*, **5**:329–359, (1996).

57. J.T. Cox and R. Durrett. Oriented percolation in dimensions $d \geq 4$: bounds and asymptotic formulas. *Math. Proc. Cambridge Philos. Soc.*, **93**:151–162, (1983).

58. J.T. Cox, R. Durrett, and E.A. Perkins. Rescaled voter models converge to super-Brownian motion. *Ann. Probab.*, **28**:185–234, (2000).

59. T. Cox, R. Durrett, and E.A. Perkins. Rescaled particle systems converging to super-Brownian motion. In M. Bramson and R. Durrett, editors, *Perplexing Problems in Probability: Festschrift in Honor of Harry Kesten*, Basel, (1999). Birkhäuser.

60. D.A. Dawson. Measure-valued Markov processes. In *Ecole d'Eté de Prob-abilités de Saint–Flour 1991. Lecture Notes in Mathematics #1541*, Berlin, (1993). Springer.

61. E. Derbez and G. Slade. Lattice trees and super-Brownian motion. *Canad. Math. Bull.*, **40**:19–38, (1997).

62. E. Derbez and G. Slade. The scaling limit of lattice trees in high dimensions. *Commun. Math. Phys.*, **193**:69–104, (1998).

63. C. Domb and M.E. Fisher. On random walks with restricted reversals. *Proc. Camb. Phil. Soc.*, **54**:48–59, (1958).

64. C. Domb and G.S. Joyce. Cluster expansion for a polymer chain. *J. Phys. C: Solid State Phys.*, **5**:956–976, (1972).

65. B. Duplantier. Polymer chains in four dimensions. *Nuclear Physics B*, **275** [FS17]:319–355, (1986).

66. B. Duplantier. Statistical mechanics of polymer networks of any topology. *J. Stat. Phys.*, **54**:581–680, (1989).

67. B. Duplantier. Renormalization and conformal invariance for polymers. In H. van Beijeren, editor, *Fundamental Problems in Statistical Mechanics VII*, pages 171–223, Amsterdam, (1990). Elsevier Science Publishers B.V.

68. R. Durrett and E.A. Perkins. Rescaled contact processes converge to super-Brownian motion in two or more dimensions. *Probab. Th. Rel. Fields*, **114**:309–399, (1999).

69. E.B. Dynkin. Representation for functionals of superprocesses by multiple sto-chastic integrals, with applications to self-intersection local times. *Astérisque*, **157–158**:147–171, (1988).

70. I.G. Enting and A.J. Guttmann. Polygons on the honeycomb lattice. *J. Phys. A: Math. Gen.*, **22**:1371–1384, (1989).

71. P. Erdős and A. Rényi. On the evolution of random graphs. *Magyar Tud. Akad. Mat. Kutató Int. Közl.*, **5**:17–61, (1960).

72. M.H. Ernst. Random walks with short memory. *J. Stat. Phys.*, **53**:191–201, (1988).

73. R. Fernández, J. Fröhlich, and A.D. Sokal. *Random Walks, Critical Phenom-ena, and Triviality in Quantum Field Theory*. Springer, Berlin, (1992).

74. M.E. Fisher and D.S. Gaunt. Ising model and self-avoiding walks on hyper-cubical lattices and "high-density" expansions. *Phys. Rev.*, **133**:A224–A239, (1964).

75. P. Flajolet and A. Odlyzko. Singularity analysis of generating functions. *SIAM J. Disc. Math.*, **3**:216–240, (1990).

76. J. Fröhlich. On the triviality of φ_d^4 theories and the approach to the critical point in $d \geq 4$ dimensions. *Nucl. Phys.*, **B200** [FS4]:281–296, (1982).

77. D.S. Gaunt. $1/d$ expansions for critical amplitudes. *J. Phys. A: Math. Gen.*, **19**:L149–L153, (1986).

78. D.S. Gaunt and H. Ruskin. Bond percolation processes in d dimensions. *J. Phys. A: Math. Gen.*, **11**:1369–1380, (1978).

79. P.G. de Gennes. Exponents for the excluded volume problem as derived by the Wilson method. *Phys. Lett.*, **A38**:339–340, (1972).

80. P.G. de Gennes. *Scaling Concepts in Polymer Physics*. Cornell University Press, Ithaca, (1979).

81. S. Golowich and J.Z. Imbrie. A new approach to the long-time behavior of self-avoiding random walks. *Ann. Phys.*, **217**:142–169, (1992).

82. D.M. Gordon. Percolation in high dimensions. *J. London Math. Soc. (2)*, **44**:373–384, (1991).

83. I.S. Gradshteyn and I.M. Ryzhik. *Table of Integrals, Series and Products.* Academic Press, New York, 4th edition, (1965).

84. A. Greven and F. den Hollander. A variational characterization of the speed of a one-dimensional self-repellent random walk. *Ann. Appl. Probab.*, **3**:1067–1099, (1993).

85. G. Grimmett. *Percolation.* Springer, Berlin, 2nd edition, (1999).

86. G. Grimmett and P. Hiemer. Directed percolation and random walk. In V. Sidoravicius, editor, *In and Out of Equilibrium*, pages 273–297. Birkhäuser, Boston, (2002).

87. G. Grimmett and D.R. Stirzaker. *Probability and Random Processes.* Oxford University Press, Oxford, 3rd edition, (2001).

88. J.M. Hammersley and D.J.A. Welsh. Further results on the rate of convergence to the connective constant of the hypercubical lattice. *Quart. J. Math. Oxford*, (2), **13**:108–110, (1962).

89. T. Hara. Mean field critical behaviour for correlation length for percolation in high dimensions. *Prob. Th. and Rel. Fields*, **86**:337–385, (1990).

90. T. Hara. Decay of correlations in nearest-neighbour self-avoiding walk, percolation, lattice trees and animals. Preprint, (2005).

91. T. Hara, R. van der Hofstad, and G. Slade. Critical two-point functions and the lace expansion for spread-out high-dimensional percolation and related models. *Ann. Probab.*, **31**:349–408, (2003).

92. T. Hara and G. Slade. The mean-field critical behaviour of percolation in high dimensions. In B. Simon, A. Truman, and I.M. Davies, editors, *Proceedings of the IXth International Congress on Mathematical Physics*, pages 450–453, Bristol and New York, (1989). Adam Hilger.

93. T. Hara and G. Slade. The triangle condition for percolation. *Bull. A.M.S.*, **21**:269–273, (1989).

94. T. Hara and G. Slade. Mean-field critical behaviour for percolation in high dimensions. *Commun. Math. Phys.*, **128**:333–391, (1990).

95. T. Hara and G. Slade. On the upper critical dimension of lattice trees and lattice animals. *J. Stat. Phys.*, **59**:1469–1510, (1990).

96. T. Hara and G. Slade. Critical behaviour of self-avoiding walk in five or more dimensions. *Bull. A.M.S.*, **25**:417–423, (1991).

97. T. Hara and G. Slade. Self-avoiding walk in five or more dimensions. I. The critical behaviour. *Commun. Math. Phys.*, **147**:101–136, (1992).

98. T. Hara and G. Slade. The lace expansion for self-avoiding walk in five or more dimensions. *Reviews in Math. Phys.*, **4**:235–327, (1992).

99. T. Hara and G. Slade. The number and size of branched polymers in high dimensions. *J. Stat. Phys.*, **67**:1009–1038, (1992).

100. T. Hara and G. Slade. Mean-field behaviour and the lace expansion. In G. Grimmett, editor, *Probability and Phase Transition*, Dordrecht, (1994). Kluwer.

101. T. Hara and G. Slade. The self-avoiding-walk and percolation critical points in high dimensions. *Combin. Probab. Comput.*, **4**:197–215, (1995).

102. T. Hara and G. Slade. Unpublished appendix to [101]. Available as paper 93-288 at http://www.ma.utexas.edu/mp_arc. (1993).

103. T. Hara and G. Slade. The incipient infinite cluster in high-dimensional percolation. *Electron. Res. Announc. Amer. Math. Soc.*, **4**:48–55, (1998). http://www.ams.org/era/.

104. T. Hara and G. Slade. The scaling limit of the incipient infinite cluster in high-dimensional percolation. I. Critical exponents. *J. Stat. Phys.*, **99**:1075–1168, (2000).

105. T. Hara and G. Slade. The scaling limit of the incipient infinite cluster in high-dimensional percolation. II. Integrated super-Brownian excursion. *J. Math. Phys.*, **41**:1244–1293, (2000).

106. T. Hara, G. Slade, and A.D. Sokal. New lower bounds on the self-avoiding-walk connective constant. *J. Stat. Phys.*, **72**:479–517, (1993). Erratum: *J. Stat. Phys.*, **78**:1187–1188, (1995).

107. T.E. Harris. Contact interactions on a lattice. *Ann. Probab.*, **2**:969–988, (1974).

108. R. van der Hofstad. The lace expansion approach to ballistic behaviour for one-dimensional weakly self-avoiding walk. *Probab. Th. Rel. Fields*, **119**:311–349, (2001).

109. R. van der Hofstad. Infinite canonical super-Brownian motion and scaling limits. To appear in *Commun. Math. Phys.*

110. R. van der Hofstad. Spread-out oriented percolation and related models above the upper critical dimension: induction and super-processes. *Ensaios Matemáticos*, **9**:91–181, (2005).

111. R. van der Hofstad, F. den Hollander, and G. Slade. The survival probability for critical spread-out oriented percolation above $4 + 1$ dimensions. I. Induction. Preprint, (2005).

112. R. van der Hofstad, F. den Hollander, and G. Slade. The survival probability for critical spread-out oriented percolation above $4 + 1$ dimensions. II. Expansion. Preprint, (2005).

113. R. van der Hofstad, F. den Hollander, and G. Slade. A new inductive approach to the lace expansion for self-avoiding walks. *Probab. Th. Rel. Fields*, **111**:253–286, (1998).

114. R. van der Hofstad, F. den Hollander, and G. Slade. Construction of the incipient infinite cluster for spread-out oriented percolation above $4 + 1$ dimensions. *Commun. Math. Phys.*, **231**:435–461, (2002).

115. R. van der Hofstad and A.A. Járai. The incipient infinite cluster for high-dimensional unoriented percolation. *J. Stat. Phys*, **114**:625–663, (2004).

116. R. van der Hofstad and W. König. A survey of one-dimensional random polymers. *J. Stat. Phys.*, **103**:915–944, (2001).

117. R. van der Hofstad and A. Sakai. Convergence of the critical finite-range contact process to super-Brownian motion above the upper critical dimension. I. The higher point functions. II. The survival probability. In preparation.

118. R. van der Hofstad and A. Sakai. Critical points for spread-out self-avoiding walk, percolation and the contact process. *Probab. Th. Rel. Fields*, **132**:438–470, (2005).

119. R. van der Hofstad and A. Sakai. Gaussian scaling for the critical spread-out contact process above the upper critical dimension. *Electron. J. Probab.*, **9**:710–769, (2004).

120. R. van der Hofstad and G. Slade. A generalised inductive approach to the lace expansion. *Probab. Th. Rel. Fields*, **122**:389–430, (2002).

121. R. van der Hofstad and G. Slade. Convergence of critical oriented percolation to super-Brownian motion above $4 + 1$ dimensions. *Ann. Inst. H. Poincaré Probab. Statist.*, **39**:415–485, (2003).

122. R. van der Hofstad and G. Slade. Asymptotic expansions in n^{-1} for percolation critical values on the n-cube and $\mathbb{Z}^n$. *Random Struct. Alg.*, **27**:331–357, (2005).

123. R. van der Hofstad and G. Slade. Expansion in n^{-1} for percolation critical values on the n-cube and $\mathbb{Z}^n$: the first three terms. To appear in *Combin. Probab. Comput.*

124. R. van der Hofstad and G. Slade. The lace expansion on a tree with application to networks of self-avoiding walks. *Adv. Appl. Math.*, **30**:471–528, (2003).

125. M. Holmes. *Convergence of lattice trees to super-Brownian motion above the critical dimension*. PhD thesis, University of British Columbia, (2005).

126. M. Holmes, A.A. Járai, A. Sakai, and G. Slade. High-dimensional graphical networks of self-avoiding walks. *Canad. J. Math.*, **56**:77–114, (2004).

127. B.D. Hughes. *Random Walks and Random Environments*, volume 1: Random Walks. Oxford University Press, Oxford, (1995).

128. B.D. Hughes. *Random Walks and Random Environments*, volume 2: Random Environments. Oxford University Press, Oxford, (1996).

129. D. Iagolnitzer and J. Magnen. Polymers in a weak random potential in dimension four: rigorous renormalization group analysis. *Commun. Math. Phys.*, **162**:85–121, (1994).

130. D. Ioffe. Ornstein-Zernike behaviour and analyticity of shapes for self-avoiding walks on Z^d. *Markov Process. Related Fields*, **4**:323–350, (1998).

131. E. J. Janse van Rensburg. On the number of trees in $\mathbf{Z}^d$. *J. Phys. A: Math. Gen.*, **25**:3523–3528, (1992).

132. E. J. Janse van Rensburg. *The Statistical Mechanics of Interacting Walks, Polygons, Animals and Vesicles*. Oxford University Press, Oxford, (2000).

133. E. J. Janse van Rensburg and N. Madras. A non-local Monte–Carlo algorithm for lattice trees. *J. Phys. A: Math. Gen.*, **25**:303–333, (1992).

134. S. Janson, T. Łuczak, and A. Ruciński. *Random Graphs*. John Wiley and Sons, New York, (2000).

135. A.A. Járai. Incipient infinite percolation clusters in $2d$. *Ann. Probab.*, **31**:444–485, (2003).

136. A.A. Járai. Invasion percolation and the incipient infinite cluster in $2d$. *Commun. Math. Phys.*, **236**:311–334, (2003).

137. I. Jensen and A.J. Guttmann. Self-avoiding polygons on the square lattice. *J. Phys. A: Math. Gen.*, **32**:4867–4876, (1999).

138. T. Kennedy. Ballistic behavior in a 1D weakly self-avoiding walk with decaying energy penalty. *J. Stat. Phys.*, **77**:565–579, (1994).

139. T. Kennedy. Conformal invariance and stochastic Loewner evolution predictions for the 2D self-avoiding walk–Monte Carlo tests. *J. Stat. Phys.*, **114**:51–78, (2004).

140. H. Kesten. On the number of self-avoiding walks. II. *J. Math. Phys.*, **5**:1128–1137, (1964).

141. H. Kesten. *Percolation Theory for Mathematicians*. Birkhäuser, Boston, (1982).

142. H. Kesten. The incipient infinite cluster in two-dimensional percolation. *Probab. Th. Rel. Fields*, **73**:369–394, (1986).

143. H. Kesten. Asymptotics in high dimensions for percolation. In G.R. Grimmett and D.J.A. Welsh, editors, *Disorder in Physical Systems*. Clarendon Press, Oxford, (1990).

144. K.M. Khanin, J.L. Lebowitz, A.E. Mazel, and Ya.G. Sinai. Self-avoiding walks in five or more dimensions: polymer expansion approach. *Russian Math. Surveys*, **50**:403–434, (1995).

145. D.J. Klein. Rigorous results for branched polymer models with excluded volume. *J. Chem. Phys.*, **75**:5186–5189, (1981).

146. W. König. A central limit theorem for a one-dimensional polymer measure. *Ann. Probab*, **24**:1012–1035, (1996).

147. G.F. Lawler. A self-avoiding random walk. *Duke Math. J.*, **47**:655–693, (1980).

148. G.F. Lawler. The infinite self-avoiding walk in high dimensions. *Ann. Probab.*, **17**:1367–1376, (1989).

149. G.F. Lawler. *Intersections of Random Walks*. Birkhäuser, Boston, (1991).

150. G.F. Lawler, O. Schramm, and W. Werner. On the scaling limit of planar self-avoiding walk. In *Fractal geometry and applications: a jubilee of Benoît Mandelbrot, Part 2*, pages 339–364, Proc. Sympos. Pure Math., 72, Part 2. Amer. Math. Soc., Providence, RI, (2004).

151. J.-F. Le Gall. The uniform random tree in a Brownian excursion. *Probab. Th. Rel. Fields*, **96**:369–383, (1993).

152. J.-F. Le Gall. *Spatial Branching Processes, Random Snakes, and Partial Differential Equations*. Birkhäuser, Basel, (1999).

153. T.M. Liggett. *Interacting Particle Systems*. Springer, New York, (1985).

154. T.M. Liggett. *Stochastic Interacting Systems: Contact, Voter and Exclusion Processes*. Springer, Berlin, (1999).

155. T.C. Lubensky and J. Isaacson. Statistics of lattice animals and dilute branched polymers. *Phys. Rev.*, **A20**:2130–2146, (1979).

156. T. Łuczak. Component behavior near the critical point of the random graph process. *Random Structures Algorithms*, **1**:287–310, (1990).

157. N. Madras. A rigorous bound on the critical exponent for the number of lattice trees, animals and polygons. *J. Stat. Phys.*, **78**:681–699, (1995).

158. N. Madras and G. Slade. *The Self-Avoiding Walk*. Birkhäuser, Boston, (1993).

159. N. Madras and A.D. Sokal. The pivot algorithm: A highly efficient Monte Carlo method for the self-avoiding walk. *J. Stat. Phys.*, **50**:109–186, (1988).

160. R. Meester and R. Roy. *Continuum Percolation*. Cambridge University Press, Cambridge, (1996).

161. M.V. Menshikov. Coincidence of critical points in percolation problems. *Soviet Mathematics, Doklady*, **33**:856–859, (1986).

162. M.V. Menshikov, S.A. Molchanov, and A.F. Sidorenko. Percolation theory and some applications. *Itogi Nauki i Tekhniki (Series of Probability Theory, Mathematical Statistics, Theoretical Cybernetics)*, **24**:53–110, (1986). English Translation: *J. of Soviet Mathematics*, **42**:1766–1810, (1988).

163. A.M. Nemirovsky, K.F. Freed, T. Ishinabe, and J.F. Douglas. End-to-end distance of a single self-interacting self-avoiding polymer chain: d^{-1} expansion. *Phys. Lett. A*, **162**:469–474, (1992).

164. A.M. Nemirovsky, K.F. Freed, T. Ishinabe, and J.F. Douglas. Marriage of exact enumeration and $1/d$ expansion methods: Lattice model of dilute polymers. *J. Stat. Phys.*, **67**:1083–1108, (1992).

165. B.G. Nguyen. Gap exponents for percolation processes with triangle condition. *J. Stat. Phys.*, **49**:235–243, (1987).

166. B.G. Nguyen and W-S. Yang. Triangle condition for oriented percolation in high dimensions. *Ann. Probab.*, **21**:1809–1844, (1993).

167. B.G. Nguyen and W-S. Yang. Gaussian limit for critical oriented percolation in high dimensions. *J. Stat. Phys.*, **78**:841–876, (1995).

168. B. Nienhuis. Exact critical exponents of the $O(n)$ models in two dimensions. *Phys. Rev. Lett.*, **49**:1062–1065, (1982).

169. B. Nienhuis. Critical behavior of two-dimensional spin models and charge asymmetry in the Coulomb gas. *J. Stat. Phys.*, **34**:731–761, (1984).

170. B. Nienhuis. Coulomb gas formulation of two-dimensional phase transitions. In C. Domb and J.L. Lebowitz, editors, *Phase Transitions and Critical Phenomena*, volume 11, New York, (1987). Academic Press.

171. J. Noonan. New upper bounds on the connective constants of self-avoiding walks. *J. Stat. Phys.*, **91**:871–888, (1998).

172. S.P. Obukhov. The problem of directed percolation. *Physica*, **101A**:145–155, (1980).

173. A.L. Owczarek and T. Prellberg. Scaling of self-avoiding walks in high dimensions. *J. Phys. A: Math. Gen.*, **34**:5773–5780, (2001).

174. A.L. Owczarek and T. Prellberg. Monte Carlo investigations of lattice models of polymer collapse in five dimensions. *Internat. J. Modern Phys. C*, **14**:621–633, (2003).

175. G. Parisi and N. Sourlas. Critical behavior of branched polymers and the Lee–Yang edge singularity. *Phys. Rev. Lett.*, **46**:871–874, (1981).

176. M.D. Penrose. Self-avoiding walks and trees in spread-out lattices. *J. Stat. Phys.*, **77**:3–15, (1994).

177. E. Perkins. Dawson–Watanabe superprocesses and measure-valued diffusions. In P.L. Bernard, editor, *Lectures on Probability Theory and Statistics. Ecole d'Eté de Probabilités de Saint–Flour XXIX-1999*, pages 125–329, Berlin, (2002). Springer. Lecture Notes in Mathematics #1781.

178. E. Perkins. Super-Brownian motion and critical spatial stochastic systems. *Canad. Math. Bull.*, **47**:280–297, (2004).

179. A. Sakai. Mean-field critical behavior for the contact process. *J. Stat. Phys.*, **104**:111–143, (2001).

180. A. Sakai. Hyperscaling inequalities for the contact process and oriented percolation. *J. Stat. Phys.*, **106**:201–211, (2002).

181. A. Sakai. Lace expansion for the Ising model. Preprint, (2005).

182. L. Serlet. Super-Brownian motion conditioned on the total mass. Preprint, (2005).

183. G. Slade. The diffusion of self-avoiding random walk in high dimensions. *Commun. Math. Phys.*, **110**:661–683, (1987).

184. G. Slade. Convergence of self-avoiding random walk to Brownian motion in high dimensions. *J. Phys. A: Math. Gen.*, **21**:L417–L420, (1988).

185. G. Slade. The scaling limit of self-avoiding random walk in high dimensions. *Ann. Probab.*, **17**:91–107, (1989).

186. G. Slade. The lace expansion and the upper critical dimension for percolation. *Lectures in Applied Mathematics*, **27**:53–63, (1991). (Mathematics of Random Media, eds. W.E. Kohler and B.S. White, A.M.S., Providence).

187. G. Slade. Bounds on the self-avoiding-walk connective constant. *Journal of Fourier Analysis and Applications*, Special Issue: Proceedings of the Conference in Honor of Jean-Pierre Kahane (Orsay, June 28 – July 3 1993):525–533, (1995).

188. G. Slade. Lattice trees, percolation and super-Brownian motion. In M. Bramson and R. Durrett, editors, *Perplexing Problems in Probability: Festschrift in Honor of Harry Kesten*, Basel, (1999). Birkhäuser.

189. G. Slade. Scaling limits and super-Brownian motion. *Notices A.M.S.*, **49**(9):1056–1067, (2002).

190. S. Smirnov. Critical percolation in the plane. I. Conformal invariance and Cardy's formula. II. Continuum scaling limit. Preprint, (2001).

191. S. Smirnov and W. Werner. Critical exponents for two-dimensional percolation. *Math. Res. Lett.*, **8**:729–744, (2001).

192. A.D. Sokal. A rigorous inequality for the specific heat of an Ising or φ^4 ferromagnet. *Phys. Lett.*, **71A**:451–453, (1979).

193. C.E. Soteros. Private communication.

194. C.E. Soteros, D.W. Sumners, and S.G. Whittington. Entanglement complexity of graphs in Z^3. *Math. Proc. Camb. Phil. Soc.*, **111**:75–91, (1992).

195. F. Spitzer. *Principles of Random Walk.* Springer, New York, 2nd edition, (1976).

196. R.P. Stanley. *Enumerative Combinatorics*, volume 2. Cambridge University Press, Cambridge, (1999).

197. D. Stauffer and A. Aharony. *Introduction to Percolation Theory.* Taylor and Francis, London, 2nd edition, (1992).

198. H. Tanemura. Critical behavior for a continuum percolation model. In S. Watanabe, M. Fukushima, Yu.V. Prohorov, and A.N. Shiryaev, editors, *Probability theory and mathematical statistics (Tokyo 1995)*, pages 485–495, River Edge, N.J., (1996). World Sci. Publishing.

199. H. Tasaki. *Stochastic geometric methods in statistical physics and field theories.* PhD thesis, University of Tokyo, (1986).

200. H. Tasaki. Hyperscaling inequalities for percolation. *Commun. Math. Phys.*, **113**:49–65, (1987).

201. H. Tasaki and T. Hara. Critical behaviour in a system of branched polymers. *Prog. Theor. Phys. Suppl.*, **92**:14–25, (1987).

202. G. Toulouse. Perspectives from the theory of phase transitions. *Nuovo Cimento*, **23B**:234–240, (1974).

203. K. Uchiyama. Green's functions for random walks on $\mathbb{Z}^N$. *Proc. London Math. Soc.*, **77**:215–240, (1998).

204. D. Ueltschi. A self-avoiding walk with attractive interactions. *Probab. Th. Rel. Fields*, **124**:189–203, (2002).

205. C. Vanderzande. *Lattice Models of Polymers.* Cambridge University Press, Cambridge, (1998).

206. W. Werner. *Random planar curves and Schramm–Loewner evolutions.* In J. Picard, editor, *Lectures on Probability Theory and Statistics. Ecole d'Eté de Probabilités de Saint-Flour XXXII-2002.* Springer, Berlin (2004). Lecture Notes in Mathematics #1840.

207. W-S. Yang and D. Klein. A note on the critical dimension for weakly self-avoiding walks. *Prob. Th. and Rel. Fields*, **79**:99–114, (1988).

208. W-S. Yang and B. Nguyen. Gaussian limit for oriented percolation in high dimensions. In G.J. Morrow and W-S. Yang, editors, *Probability Models in Mathematical Physics*, Teaneck, NJ, (1991). World Scientific.

209. W-S. Yang and A. Zeleke. Expansion methods and scaling limits above critical dimensions. *Taiwanese J. Math.*, **3**:425–474, (1999).

210. W-S. Yang and Y. Zhang. A note on differentiability of the cluster density for independent percolation in high dimensions. *J. Stat. Phys.*, **66**:1123–1138, (1992).

211. D. Zeilberger. The abstract lace expansion. *Adv. Appl. Math.*, **19**:355–359, (1997).

Index

List of participants

Lecturers

CERF Raphaël	Univ. Paris-Sud, Orsay, F
LYONS Terry	Univ. Oxford, UK
SLADE Gordon	Univ. British Columbia, Vancouver, Canada

Participants

ASSELAH Amine	Univ. Provence, Marseille, F
AUTRET Solenn	Univ. Paul Sabatier, Toulouse, F
BAILLEUL Ismaël	Univ. Paris-Sud, Orsay, F
BAUDOIN Fabrice	Univ. Paul Sabatier, Toulouse, F
BEGYN Arnaud	Univ. Paul Sabatier, Toulouse, F
BEN-ARI Iddo	Technion Inst. Technology, Haifa, Israel
BERARD Jean	Univ. Lyon 1, F
BLACHE Fabrice	Univ. Blaise Pascal, Clermont-Ferrand, F
BROMAN Erik	Chalmers Univ. Techn., Gothenburg, Sweden
BROUTTELANDE Christophe	Univ. Paul Sabatier, Toulouse, F
BRYC Wlodzimierz	Univ. Cincinnati, USA
CARUANA Michael	Univ. Oxford, UK
CHIVORET Sebastien	Univ. Michigan, Ann Arbor, USA
COUPIER David	Univ. Paris 5, F
CROYDON David	Univ. Oxford, UK
DE CARVALHO BEZERRA S.	Univ. Henri Poincaré, Nancy, F
DE TILIERE Béatrice	Univ. Paris-Sud, Orsay, F
DELARUE François	Univ. Paris 7, F
DEVAUX Vincent	Univ. Rouen, F
DUNLOP François	Univ. Cergy-Pontoise, F
DUQUESNE Thomas	Univ. Paris-Sud, Orsay, F
FERAL Delphine	Univ. Paul Sabatier, Toulouse, F

FILLIGER Roger	EPFL, Lausanne, Switzerland
GARET Olivier	Univ. Orléans, F
GAUTIER Eric	Univ. Rennes 1 & INSEE, F
GOERGEN Laurent	ETH Zurich, Switzerland
GOUERE Jean-Baptiste	Univ. Claude Bernard, Lyon, F
GOURCY Mathieu	Univ. Blaise Pascal, Clermont-Ferrand, F
HOLROYD Alexander	Univ. British Columbia, Vancouver, Canada
ISHIKAWA Yasushi	Univ. Ehime, Matsuyama, Japan
JAKUBOWICZ Jérémie	ENS Cachan, F
JOULIN Aldéric	Univ. La Rochelle, F
KASPRZYK Arkadiusz	Univ. Wroclaw, Poland
KOVCHEGOV Yevgeniy	UCLA, Los Angeles, USA
KURT Noemi	Univ. Zurich, Switzerland
LACAUX Céline	Univ. Paul Sabatier, Toulouse, F
LACHAUD Béatrice	Univ. Paris 5, F
LE GALL Jean-Francçois	ENS Paris, F
LE JAN Yves	Univ. Paris-Sud, Orsay, F
LEI Liangzhen	Univ. Blaise Pascal, Clermont-Ferrand, F
LEVY Thierry	ENS Paris, F
LUCZAK Malwina	London School of Economics, UK
MARCHAND Régine	Univ. Henri Poincaré, Nancy, F
MARDIN Arif	TELECOM-INT, Evry, F
MARTIN James	Univ. Paris 7, F
MARTY Renaud	Univ. Paul Sabatier, Toulouse, F
MERLE Mathieu	ENS Paris, F
MESSIKH Reda Jürg	Univ. Paris-Sud, Orsay, F
MOCIOALCA Oana	Purdue Univ., West Lafayette, USA
NIEDERHAUSEN Meike	Purdue Univ., West Lafayette, USA
NINOMIYA Syoiti	Tokyo Instit. Technology, Japan
NUALART David	Univ. Barcelona, Spain
PICARD Jean	Univ. Blaise Pascal, Clermont-Ferrand, F
PUDLO Pierre	Univ. Claude Bernard, Lyon, F
RIVIERE Olivier	Univ. Paris 5, F
ROUSSET Mathias	Univ. Paul Sabatier, Toulouse, F
ROUX Daniel	Univ. Blaise Pascal, Clermont-Ferrand, F
SAINT LOUBERT BIE Erwan	Univ. Blaise Pascal, Clermont-Ferrand, F
SAVONA Catherine	Univ. Blaise Pascal, Clermont-Ferrand, F
SERLET Laurent	Univ. Paris 5, F
THOMANN Philipp	Univ. Zurich, Switzerland
TORRECILLA Iván	Univ. Barcelona, Spain
TRASHORRAS Jose	Univ. Warwick, Coventry, UK
TURNER Amanda	Univ. Cambridge, UK
TYKESSON Johan	Chalmers Univ. Techn., Göteborg, Sweden
VIGNAUD Yvon	CPT, Marseille, F
WEILL Mathilde	ENS Paris, F
WINKEL Matthias	Univ. Oxford, UK
YU Yuhua	Purdue Univ., West Lafayette, USA

List of short lectures

Lecture Notes in Mathematics

For information about earlier volumes
please contact your bookseller or Springer
LNM Online archive: springerlink.com

Vol. 1788: A. Vasil'ev, Moduli of Families of Curves for Conformal and Quasiconformal Mappings (2002)

Vol. 1789: Y. Sommerhäuser, Yetter-Drinfel'd Hopf algebras over groups of prime order (2002)

Vol. 1790: X. Zhan, Matrix Inequalities (2002)

Vol. 1791: M. Knebusch, D. Zhang, Manis Valuations and Prüfer Extensions I: A new Chapter in Commutative Algebra (2002)

Vol. 1792: D. D. Ang, R. Gorenflo, V. K. Le, D. D. Trong, Moment Theory and Some Inverse Problems in Potential Theory and Heat Conduction (2002)

Vol. 1793: J. Cortés Monforte, Geometric, Control and Numerical Aspects of Nonholonomic Systems (2002)

Vol. 1794: N. Pytheas Fogg, Substitution in Dynamics, Arithmetics and Combinatorics. Editors: V. Berthé, S. Ferenczi, C. Mauduit, A. Siegel (2002)

Vol. 1795: H. Li, Filtered-Graded Transfer in Using Noncommutative Gröbner Bases (2002)

Vol. 1796: J.M. Melenk, hp-Finite Element Methods for Singular Perturbations (2002)

Vol. 1797: B. Schmidt, Characters and Cyclotomic Fields in Finite Geometry (2002)

Vol. 1798: W.M. Oliva, Geometric Mechanics (2002)

Vol. 1799: H. Pajot, Analytic Capacity, Rectifiability, Menger Curvature and the Cauchy Integral (2002)

Vol. 1800: O. Gabber, L. Ramero, Almost Ring Theory (2003)

Vol. 1801: J. Azéma, M. Émery, M. Ledoux, M. Yor (Eds.), Séminaire de Probabilités XXXVI (2003)

Vol. 1802: V. Capasso, E. Merzbach, B.G. Ivanoff, M. Dozzi, R. Dalang, T. Mountford, Topics in Spatial Stochastic Processes. Martina Franca, Italy 2001. Editor: E. Merzbach (2003)

Vol. 1803: G. Dolzmann, Variational Methods for Crystalline Microstructure – Analysis and Computation (2003)

Vol. 1804: I. Cherednik, Ya. Markov, R. Howe, G. Lusztig, Iwahori-Hecke Algebras and their Representation Theory. Martina Franca, Italy 1999. Editors: V. Baldoni, D. Barbasch (2003)

Vol. 1805: F. Cao, Geometric Curve Evolution and Image Processing (2003)

Vol. 1806: H. Broer, I. Hoveijn. G. Lunther, G. Vegter, Bifurcations in Hamiltonian Systems. Computing Singularities by Gröbner Bases (2003)

Vol. 1807: V. D. Milman, G. Schechtman (Eds.), Geometric Aspects of Functional Analysis. Israel Seminar 2000-2002 (2003)

Vol. 1808: W. Schindler, Measures with Symmetry Properties (2003)

Vol. 1809: O. Steinbach, Stability Estimates for Hybrid Coupled Domain Decomposition Methods (2003)

Vol. 1810: J. Wengenroth, Derived Functors in Functional Analysis (2003)

Vol. 1811: J. Stevens, Deformations of Singularities (2003)

Vol. 1812: L. Ambrosio, K. Deckelnick, G. Dziuk, M. Mimura, V. A. Solonnikov, H. M. Soner, Mathematical Aspects of Evolving Interfaces. Madeira, Funchal, Portugal 2000. Editors: P. Colli, J. F. Rodrigues (2003)

Vol. 1813: L. Ambrosio, L. A. Caffarelli, Y. Brenier, G. Buttazzo, C. Villani, Optimal Transportation and its Applications. Martina Franca, Italy 2001. Editors: L. A. Caffarelli, S. Salsa (2003)

Vol. 1814: P. Bank, F. Baudoin, H. Föllmer, L.C.G. Rogers, M. Soner, N. Touzi, Paris-Princeton Lectures on Mathematical Finance 2002 (2003)

Vol. 1815: A. M. Vershik (Ed.), Asymptotic Combinatorics with Applications to Mathematical Physics. St. Petersburg, Russia 2001 (2003)

Vol. 1816: S. Albeverio, W. Schachermayer, M. Talagrand, Lectures on Probability Theory and Statistics. Ecole d'Eté de Probabilités de Saint-Flour XXX-2000. Editor: P. Bernard (2003)

Vol. 1817: E. Koelink, W. Van Assche(Eds.), Orthogonal Polynomials and Special Functions. Leuven 2002 (2003)

Vol. 1818: M. Bildhauer, Convex Variational Problems with Linear, nearly Linear and/or Anisotropic Growth Conditions (2003)

Vol. 1819: D. Masser, Yu. V. Nesterenko, H. P. Schlickewei, W. M. Schmidt, M. Waldschmidt, Diophantine Approximation. Cetraro, Italy 2000. Editors: F. Amoroso, U. Zannier (2003)

Vol. 1820: F. Hiai, H. Kosaki, Means of Hilbert Space Operators (2003)

Vol. 1821: S. Teufel, Adiabatic Perturbation Theory in Quantum Dynamics (2003)

Vol. 1822: S.-N. Chow, R. Conti, R. Johnson, J. Mallet-Paret, R. Nussbaum, Dynamical Systems. Cetraro, Italy 2000. Editors: J. W. Macki, P. Zecca (2003)

Vol. 1823: A. M. Anile, W. Allegretto, C. Ringhofer, Mathematical Problems in Semiconductor Physics. Cetraro, Italy 1998. Editor: A. M. Anile (2003)

Vol. 1824: J. A. Navarro González, J. B. Sancho de Salas, $\mathcal{C}^\infty$ – Differentiable Spaces (2003)

Vol. 1825: J. H. Bramble, A. Cohen, W. Dahmen, Multiscale Problems and Methods in Numerical Simulations, Martina Franca, Italy 2001. Editor: C. Canuto (2003)

Vol. 1826: K. Dohmen, Improved Bonferroni Inequalities via Abstract Tubes. Inequalities and Identities of Inclusion-Exclusion Type. VIII, 113 p, 2003.

Vol. 1827: K. M. Pilgrim, Combinations of Complex Dynamical Systems. IX, 118 p, 2003.

Vol. 1828: D. J. Green, Gröbner Bases and the Computation of Group Cohomology. XII, 138 p, 2003.

Vol. 1829: E. Altman, B. Gaujal, A. Hordijk, Discrete-Event Control of Stochastic Networks: Multimodularity and Regularity. XIV, 313 p, 2003.

Vol. 1830: M. I. Gil', Operator Functions and Localization of Spectra. XIV, 256 p, 2003.

Vol. 1831: A. Connes, J. Cuntz, E. Guentner, N. Higson, J. E. Kaminker, Noncommutative Geometry, Martina Franca, Italy 2002. Editors: S. Doplicher, L. Longo (2004)

Vol. 1832: J. Azéma, M. Émery, M. Ledoux, M. Yor (Eds.), Séminaire de Probabilités XXXVII (2003)

Vol. 1833: D.-Q. Jiang, M. Qian, M.-P. Qian, Mathematical Theory of Nonequilibrium Steady States. On the Frontier of Probability and Dynamical Systems. IX, 280 p, 2004.

Vol. 1834: Yo. Yomdin, G. Comte, Tame Geometry with Application in Smooth Analysis. VIII, 186 p, 2004.

Vol. 1835: O.T. Izhboldin, B. Kahn, N.A. Karpenko, A. Vishik, Geometric Methods in the Algebraic Theory of Quadratic Forms. Summer School, Lens, 2000. Editor: J.-P. Tignol (2004)

Vol. 1836: C. Năstăsescu, F. Van Oystaeyen, Methods of Graded Rings. XIII, 304 p, 2004.

Vol. 1837: S. Tavaré, O. Zeitouni, Lectures on Probability Theory and Statistics. Ecole d'Eté de Probabilités de Saint-Flour XXXI-2001. Editor: J. Picard (2004)

Vol. 1838: A.J. Ganesh, N.W. O'Connell, D.J. Wischik, Big Queues. XII, 254 p, 2004.

Vol. 1839: R. Gohm, Noncommutative Stationary Processes. VIII, 170 p, 2004.

Vol. 1840: B. Tsirelson, W. Werner, Lectures on Probability Theory and Statistics. Ecole d'Eté de Probabilités de Saint-Flour XXXII-2002. Editor: J. Picard (2004)

Vol. 1841: W. Reichel, Uniqueness Theorems for Variational Problems by the Method of Transformation Groups (2004)

Vol. 1842: T. Johnsen, A.L. Knutsen, K3 Projective Models in Scrolls (2004)

Vol. 1843: B. Jefferies, Spectral Properties of Noncommuting Operators (2004)

Vol. 1844: K.F. Siburg, The Principle of Least Action in Geometry and Dynamics (2004)

Vol. 1845: Min Ho Lee, Mixed Automorphic Forms, Torus Bundles, and Jacobi Forms (2004)

Vol. 1846: H. Ammari, H. Kang, Reconstruction of Small Inhomogeneities from Boundary Measurements (2004)

Vol. 1847: T.R. Bielecki, T. Björk, M. Jeanblanc, M. Rutkowski, J.A. Scheinkman, W. Xiong, Paris-Princeton Lectures on Mathematical Finance 2003 (2004)

Vol. 1848: M. Abate, J. E. Fornaess, X. Huang, J. P. Rosay, A. Tumanov, Real Methods in Complex and CR Geometry, Martina Franca, Italy 2002. Editors: D. Zaitsev, G. Zampieri (2004)

Vol. 1849: Martin L. Brown, Heegner Modules and Elliptic Curves (2004)

Vol. 1850: V. D. Milman, G. Schechtman (Eds.), Geometric Aspects of Functional Analysis. Israel Seminar 2002-2003 (2004)

Vol. 1851: O. Catoni, Statistical Learning Theory and Stochastic Optimization (2004)

Vol. 1852: A.S. Kechris, B.D. Miller, Topics in Orbit Equivalence (2004)

Vol. 1853: Ch. Favre, M. Jonsson, The Valuative Tree (2004)

Vol. 1854: O. Saeki, Topology of Singular Fibers of Differential Maps (2004)

Vol. 1855: G. Da Prato, P.C. Kunstmann, I. Lasiecka, A. Lunardi, R. Schnaubelt, L. Weis, Functional Analytic Methods for Evolution Equations. Editors: M. Iannelli, R. Nagel, S. Piazzera (2004)

Vol. 1856: K. Back, T.R. Bielecki, C. Hipp, S. Peng, W. Schachermayer, Stochastic Methods in Finance, Bressanone/Brixen, Italy, 2003. Editors: M. Fritelli, W. Runggaldier (2004)

Vol. 1857: M. Émery, M. Ledoux, M. Yor (Eds.), Séminaire de Probabilités XXXVIII (2005)

Vol. 1858: A.S. Cherny, H.-J. Engelbert, Singular Stochastic Differential Equations (2005)

Vol. 1859: E. Letellier, Fourier Transforms of Invariant Functions on Finite Reductive Lie Algebras (2005)

Vol. 1860: A. Borisyuk, G.B. Ermentrout, A. Friedman, D. Terman, Tutorials in Mathematical Biosciences I. Mathematical Neurosciences (2005)

Vol. 1861: G. Benettin, J. Henrard, S. Kuksin, Hamiltonian Dynamics – Theory and Applications, Cetraro, Italy, 1999. Editor: A. Giorgilli (2005)

Vol. 1862: B. Helffer, F. Nier, Hypoelliptic Estimates and Spectral Theory for Fokker-Planck Operators and Witten Laplacians (2005)

Vol. 1863: H. Fürh, Abstract Harmonic Analysis of Continuous Wavelet Transforms (2005)

Vol. 1864: K. Efstathiou, Metamorphoses of Hamiltonian Systems with Symmetries (2005)

Vol. 1865: D. Applebaum, B.V. R. Bhat, J. Kustermans, J. M. Lindsay, Quantum Independent Increment Processes I. From Classical Probability to Quantum Stochastic Calculus. Editors: M. Schürmann, U. Franz (2005)

Vol. 1866: O.E. Barndorff-Nielsen, U. Franz, R. Gohm, B. Kümmerer, S. Thorbjønsen, Quantum Independent Increment Processes II. Structure of Quantum Lévy Processes, Classical Probability, and Physics. Editors: M. Schürmann, U. Franz, (2005)

Vol. 1867: J. Sneyd (Ed.), Tutorials in Mathematical Biosciences II. Mathematical Modeling of Calcium Dynamics and Signal Transduction. (2005)

Vol. 1868: J. Jorgenson, S. Lang, $Pos_n(R)$ and Eisenstein Sereies. (2005)

Vol. 1869: A. Dembo, T. Funaki, Lectures on Probability Theory and Statistics. Ecole d'Eté de Probabilités de Saint-Flour XXXIII-2003. Editor: J. Picard (2005)

Vol. 1870: V.I. Gurariy, W. Lusky, Geometry of Müntz Spaces and Related Questions. (2005)

Vol. 1871: P. Constantin, G. Gallavotti, A.V. Kazhikhov, Y. Meyer, S. Ukai, Mathematical Foundation of Turbulent Viscous Flows, Martina Franca, Italy, 2003. Editors: M. Cannone, T. Miyakawa (2006)

Vol. 1872: A. Friedman (Ed.), Tutorials in Mathematical Biosciences III. Cell Cycle, Proliferation, and Cancer (2006)

Vol. 1873: R. Mansuy, M. Yor, Random Times and Enlargements of Filtrations in a Brownian Setting (2006)

Vol. 1874: M. Yor, M. Émery (Eds.), In Memoriam Paul-André Meyer - Séminaire de Probabilités XXXIX (2006)

Vol. 1875: J. Pitman, Combinatorial Stochastic Processes. Ecole d'Eté de Probabilités de Saint-Flour XXXII-2002. Editor: J. Picard (2006)

Vol. 1876: H. Herrlich, Axiom of Choice (2006)

Vol. 1877: J. Steuding, Value Distributions of L-Functions (2006)

Vol. 1878: R. Cerf, The Wulff Crystal in Ising and Percolation Models, Ecole d'Eté de Probabilités de Saint-Flour XXXIV-2004. Editor: Jean Picard (2006)

Vol. 1879: G. Slade, The Lace Expansion and its Applications, Ecole d'Eté de Probabilités de Saint-Flour XXXIV-2004. Editor: Jean Picard (2006)

Vol. 1880: S. Attal, A. Joye, C.-A. Pillet, Open Quantum Systems I, The Hamiltonian Approach (2006)

Vol. 1881: S. Attal, A. Joye, C.-A. Pillet, Open Quantum Systems II, The Markovian Approach (2006)

Vol. 1882: S. Attal, A. Joye, C.-A. Pillet, Open Quantum Systems III, Recent Developments (2006)

Vol. 1883: W. Van Assche, F. Marcellàn (Eds.), Orthogonal Polynomials and Special Functions, Computation and Application (2006)

Vol. 1884: N. Hayashi, E.I. Kaikina, P.I. Naumkin, I.A. Shishmarev, Asymptotics for Dissipative Nonlinear Equations (2006)

Vol. 1885: A. Telcs, The Art of Random Walks (2006)

Recent Reprints and New Editions

Vol. 1471: M. Courtieu, A.A. Panchishkin, Non-Archimedean L-Functions and Arithmetical Siegel Modular Forms. – Second Edition (2003)

Vol. 1618: G. Pisier, Similarity Problems and Completely Bounded Maps. 1995 – Second, Expanded Edition (2001)

Vol. 1629: J.D. Moore, Lectures on Seiberg-Witten Invariants. 1997 – Second Edition (2001)

Vol. 1638: P. Vanhaecke, Integrable Systems in the realm of Algebraic Geometry. 1996 – Second Edition (2001)

Vol. 1702: J. Ma, J. Yong, Forward-Backward Stochastic Differential Equations and their Applications. 1999. – Corrected 3rd printing (2005)

4. Careful preparation of the manuscripts will help keep production time short besides ensuring satisfactory appearance of the finished book in print and online. After acceptance of the manuscript authors will be asked to prepare the final LaTeX source files (and also the corresponding dvi-, pdf- or zipped ps-file) together with the final printout made from these files. The LaTeX source files are essential for producing the full-text online version of the book (see http://www.springerlink.com/openurl.asp?genre=journal&issn=0075-8434 for the existing online volumes of LNM).

   The actual production of a Lecture Notes volume takes approximately 8 weeks.

5. Authors receive a total of 50 free copies of their volume, but no royalties. They are entitled to a discount of 33.3 % on the price of Springer books purchased for their personal use, if ordering directly from Springer.

6. Commitment to publish is made by letter of intent rather than by signing a formal contract. Springer-Verlag secures the copyright for each volume. Authors are free to reuse material contained in their LNM volumes in later publications: A brief written (or e-mail) request for formal permission is sufficient.

Addresses:

Professor J.-M. Morel, CMLA,
École Normale Supérieure de Cachan,
61 Avenue du Président Wilson, 94235 Cachan Cedex, France
E-mail: Jean-Michel.Morel@cmla.ens-cachan.fr

Professor F. Takens, Mathematisch Instituut,
Rijksuniversiteit Groningen, Postbus 800,
9700 AV Groningen, The Netherlands
E-mail: F.Takens@math.rug.nl

Professor B. Teissier, Institut Mathématique de Jussieu,
UMR 7586 du CNRS, Équipe "Géométrie et Dynamique",
175 rue du Chevaleret
75013 Paris, France
E-mail: teissier@math.jussieu.fr

For the "Mathematical Biosciences Subseries" of LNM:

Professor P. K. Maini, Center for Mathematical Biology,
Mathematical Institute, 24-29 St Giles,
Oxford OX1 3LP, UK
E-mail : maini@maths.ox.ac.uk

Springer, Mathematics Editorial, Tiergartenstr. 17,
69121 Heidelberg, Germany,
Tel.: +49 (6221) 487-8410
Fax: +49 (6221) 487-8355
E-mail: lnm@springer-sbm.com